AF796256

Soil Biology

Volume 20

Series Editor
Ajit Varma, Amity Institute of Microbial Technology,
Amity University Uttar Pradesh, Noida, UP, India

Volumes published in the series
For further volumes: http://www.springer.com/series/5138

Soil Heavy Metals (Vol. 19)
I. Sherameti, A. Varma

Symbiotic Fungi (Vol. 18)
A. Varma, A.C. Kharkwal

Advances in Applied Bioremediation (Vol. 17)
A. Singh, R.C. Kuhad, O.P. Ward (Eds.)

Permafrost Soils (Vol. 16)
R. Margesin (Ed.)

Molecular Mechanisms of Plant and Microbe Coexistence (Vol. 15)
C.S. Nautiyal, P. Dion (Eds.)

Secondary Metabolites in Soil Ecology (Vol. 14)
P. Karlovsky (Ed.)

Microbiology of Extreme Soils (Vol. 13)
P. Dion, C.S. Nautiyal (Eds.)

Microbial Siderophores (Vol. 12)
A. Varma, S. Chincholkar (Eds.)

Advanced Techniques in Soil Microbiology (Vol. 11)
A. Varma, R. Oelmüller (Eds.)

Nutrient Cycling in Terrestrial Ecosystems (Vol. 10)
P. Marschner, Z. Rengel (Eds.)

Microbial Root Endophytes (Vol. 9)
B.J.E. Schulz, C.J.C. Boyle, T.N. Sieber (Eds.)

Nucleic Acids and Proteins in Soil (Vol. 8)
P. Nannipieri, K. Smalla (Eds.)

Microbial Activity in the Rhizosphere (Vol. 7)
K.G. Mukerji, C. Manoharachary, J. Singh (Eds.)

Intestinal Microorganisms of Termites and Other Invertebrates (Vol. 6)
H. König, A. Varma (Eds.)

Manual for Soil Analysis – Monitoring and Assessing Soil Bioremediation (Vol. 5)
R. Margesin, F. Schinner (Eds.)

In Vitro Culture of Mycorrhizas (Vol. 4)
S. Declerck, D.-G. Strullu, J.A. Fortin (Eds.)

Microorganisms in Soils: Roles in Genesis and Functions (Vol. 3)
F. Buscot, A. Varma (Eds.)

Athanasios P. Dedousis • Thomas Bartzanas
Editors

Soil Engineering

Springer

Editors
Dr. Athanasios P. Dedousis
Dr. Thomas Bartzanas
Center for Research & Technology – Thessaly
Institute of Technology and Management of Agricultural Eco-Systems
Technology Park of Thessaly
Volos 38500
Greece
dedousis@cereteth.gr
bartzanas@cereteth.gr

Soil Biology ISSN: 1613-3382
ISBN 978-3-642-03680-4 e-ISBN 978-3-642-03681-1
DOI 10.1007/978-3-642-03681-1
Springer Heidelberg Dordrecht London New York

Library of Congress Control Number: 2009933109

© Springer-Verlag Berlin Heidelberg 2010
This work is subject to copyright. All rights are reserved, whether the whole or part of the material is concerned, specifically the rights of translation, reprinting, reuse of illustrations, recitation, broadcasting, reproduction on microfilm or in any other way, and storage in data banks. Duplication of this publication or parts thereof is permitted only under the provisions of the German Copyright Law of September 9, 1965, in its current version, and permission for use must always be obtained from Springer. Violations are liable to prosecution under the German Copyright Law.
The use of general descriptive names, registered names, trademarks, etc. in this publication does not imply, even in the absence of a specific statement, that such names are exempt from the relevant protective laws and regulations and therefore free for general use.

Cover design: SPI Publisher Services

Printed on acid-free paper

Springer is part of Springer Science+Business Media (www.springer.com)

Preface

The agricultural world has changed significantly. In recent years there has been an increasing interest in securing the sustainability of soil by preventing it from permanent irreversible damage. The excessive use of heavy machinery, waste disposal, the use of agrochemicals and the unconventional use of soil cultivation methods has led to a series of problems forcing engineers to find solutions in these difficult areas, such as soil compaction, waste management, controlled traffic farming, optimisation of tillage tools, mechanical weed control and the use of robotics in agriculture, in order to reduce soil degradation.

This volume in the *Soil Biology* series on *Soil Engineering* is an attempt to highlight some of the aforementioned issues that have to be solved by agricultural engineers in order to ensure the sustainability of soil.

Soil movement results from man's attempts to change prevailing soil conditions into those that are more suitable, or to use soil for support and locomotion of vehicles. As the use of agricultural and forestry machinery has increased in recent years in order to increase productivity, due to the current economic situation, soil–machine interactions have changed significantly in both tillage and traction. Machinery is getting larger and heavier and threatens soils with compaction, affecting air water and nutrient movement and resulting in reduced crop production. A selection of papers in this book gives the state of the art in soil compaction.

Today the agricultural sector requires non-chemical weed control that ensures food safety without degradation of soil and water. Consumers demand high quality food products and pay special attention to food safety. Through the technical development of mechanisms for physical weed control, it might be possible to control weeds in a way that meets consumer and environmental demands.

Waste management is a vital issue in modern agriculture as volumes of waste continue to rise, leading to increased environmental risks. Application of waste to agricultural land constitutes a low-cost disposal option and can be of benefit to the soil.

Autonomous vehicles have been widely used in industrial production and warehouses, where a controlled environment can be guaranteed. In agriculture, research into driverless vehicles has always been a dream, but serious research started in the early 1960s. Possible applications for the use of robotics in agriculture are presented here, targeting soil sustainability and cost reduction.

Soil Engineering will be of great value to engineers and researchers working in the agricultural engineering section, and to postgraduate students.

The editors would like to thank the authors for their cooperation, Dr. Jutta Lindenborn from Springer for her great support during the preparation of the book, and Professor Ajit Varma, Editor of the *Soil Biology* series.

Volos, July 2009 Thanos Dedousis and Thomas Bartzanas

Contents

Contributors

Dirk Ansorge CLAAS Selbstfahrende Erntemaschinen GmbH, Münsterstr. 33, 33428 Harsewinkel, Germany, Dirk.ansorge@claas.com

Diogenes L. Antille School of Applied Sciences, Cranfield University, Bedfordshire, MK43 0AL, UK

Johan Arvidsson Department of Soil & Environment, SLU, P.O. Box 7014, SE-75007 Uppsala, Sweden, johan.arvidsson@mark.slu.se

Barbora Badalíková Research Institute for Fodder Crops, Ltd., Zahradní 1, 664 41 Troubsko, Czech Republic, badalikova@vupt.cz

S. Blackmore Center for Research and Technology, Thessaly (CERETETH), 1st Industrial zone, 38500 Volos, Greece

W.C.T. Chamen CTF Europe Ltd, Bedfordshire, UK

De Baerdemaeker, J Division of Mechatronics, Biostatistics and Sensors, Faculty of Bioscience Engineering, Kasteelpark Arenberg 30, B-3001 Heverlee, Belgium

Athanasios P. Dedousis Centre for Research and Technology, Institute of Technology and Management of Agricultural Eco-Systems, Thessaly, Greece, dedousis@cereteth.gr

Roberto Fanigliulo CRA-ING Agricultural Research Council, Agricultural Engineering Research Unit Via della Pascolare 16, 00016 Monterotondo (Rome), Italy, roberto.fanigliulo@entecra.it

László Fenyvesi Hungarian Institute of Agricultural Engineering, H-2100 Gödöllő, Tessedik str. 4, Hungary, fenyvesi@fvmmi.hu

S. Fountas Laboratory of Farm Mechanisation, Department of Agriculture, Crop Production and Rural Environment, University of Thessaly, Fytoko Street, N. Ionia, 38446 Magnesia, Greece, sfountas@uth.gr

T.A. Gemtos Laboratory of Farm Mechanisation, Department of Agriculture, Crop Production and Rural Environment, University of Thessaly, Fytoko Street, N. Ionia, 38446 Magnesia, Greece

Richard John Godwin Cranfield University, MK43 0AL, Cranfield, Bedfordshire, UK

Hans W. Griepentrog University of Copenhagen, Institute of Agriculture and Ecology, Copenhagen, Denmark

Karl Heinrich Hartge (†) Institute of Soil Science, Leibniz Universität Hannover Herrenhaeuser Str. 2, 30419 Hannover, Germany

Gunnar Hoefer Institute of Soil Science, Leibniz Universität Hannover Herrenhaeuser Str. 2, 30419 Hannover, Germany, hoefer@ifbk.uni-hannover.de

Zoltán Hudoba Hungarian Institute of Agricultural Engineering, H-2100 Gödöllő, Tessedik str. 4, Hungary, hudoba@fvmmi.hu

R. Karoui Unité de Recherches Typicité des Produits Alimentaires, ENITA Clermont Ferrand, Marmilhat, BP35, 63370 Lempdes, France

Thomas Keller SLU, Department of Soil & Environment, Box 7014, SE-75007 Uppsala, Sweden, thomas.keller@mark.slu.se

Maria I. Kokkora Technological Research Centre of Thessaly, T.E.I. of Larissa, 411 10, Larissa, Greece, mkokkora@trc-thessaly.gr

Demmel Markus Bavarian State Research Center for Agriculture, Institute for Agricultural Engineering and Animal Husbandry, Voettinger Str. 36, 85354 Freising, Germany

A.M. Mouazen Natural Resources Department, Cranfield University, MK43 OAL, United Kingdom, a.mouazen@cranfield.ac.uk

Daniele Pochi CRA-ING Agricultural Research Council, Agricultural Engineering Research Unit Via della Pascolare 16, 00016 Monterotondo (Rome) Italy, daniele.pochi@entecra.it

H. Ramon Division of Mechatronics, Biostatistics and Sensors, Faculty of Bioscience Engineering, Kasteelpark Arenberg 30, B-3001 Heverlee, Belgium

Brandhuber Robert Bavarian State Research Center for Agriculture, Institute for Agroecology, Organic Farming and Soil Protection, Voettinger Str. 38, 85354 Freising, Germany

Geischeder Rupert HARAIN Engineering GmbH, Oberharthausener Str. 14, 94333 Geiselhoering, Germany, rupert.geischeder@harain.com

J.N. Tullberg CTF Solutions, 8 Hakea Crescent, Chapel Hill, Queensland 4069, Australia

Sean F. Tyrrel Cranfield University, School of Applied Sciences, Bedfordshire, MK43 0AL, UK

G.D. Vermeulen Plant Research International, P.O. Box 616, 6700 AP Wageningen, The Netherlands, bert.vermeulen@wur.nl

Part I
Soil Tillage

Chapter 1
Draught Requirement During Tillage

Johan Arvidsson and Thomas Keller

1.1 Introduction

The main reasons for tillage are soil loosening, weed control, incorporation of crop residues and manure, and creation of a seedbed with good conditions for germination and early establishment. A distinction is generally made between primary and secondary tillage, with secondary tillage involving the creation of a seedbed.

From a soil mechanical point of view, tillage can mainly be seen as two processes: the break-up of soil and the fragmentation of soil. They occur simultaneously, although the break-up of soil takes place mainly during primary tillage, while secondary tillage (seedbed preparation) primarily aims at soil fragmentation. Draught or energy requirement can also be studied in terms of break-up or fragmentation. Most research efforts have been devoted to the break-up of soil. In many cases draught has been studied without considering the final outcome of tillage, i.e. soil fragmentation measured, for example, in terms of aggregate size distribution.

1.2 Basic Concepts

Tillage tools can be divided into four general types: tines, plough bodies, discs and rollers. In the case of a tine, the loosening effect reaches considerably further than the width of the tine body, while with a plough body the loosening effect occurs mainly within the body width (Koolen and Kuipers 1983). The most common plough body, the mouldboard plough, has distinct features and an asymmetrical

J. Arvidsson (✉) and T. Keller
Department of Soil & Environment, SLU, P.O. Box 7014, Uppsala, SE-75007, Sweden
e-mail: johan.arvidsson@mark.slu.se

A.P. Dedousis and T. Bartzanas (eds.), *Soil Engineering*, Soil Biology 20,
DOI 10.1007/978-3-642-03681-1_1, © Springer-Verlag Berlin Heidelberg 2010

shape, which makes analysis of soil break-up much more complicated than for the tine. Most tillage implements are passive, meaning that the tools are simply pulled through the soil without any extra energy input. However, they can also be active, so that energy is transferred to them for active motion, usually by the power take-off of a tractor. Active tillage implements are not further discussed here.

Most of the research on draught requirement has concentrated on tines. Simple straight tines can be characterised by three properties: the width of the tine, the working depth and the rake angle, which is the inclination between the tine and the soil in the direction of travel (Fig. 1.1). The draught requirement for a tillage tool or implement can be specified in different ways:

Draught = Force (in newtons, N) required to pull the implement through the soil.
Specific draught (or specific resistance) = Draught divided by the cross-sectional area of soil worked by the implement (Nm^{-2}) (Fig. 1.2).
Total draught = Draught per metre working width (Nm^{-1})

Most research has considered only the draught force, but to determine the efficiency of an implement in soil break-up, the specific draught is more useful. From a practical point of view the total draught is important since it is directly correlated with the fuel consumption per hectare.

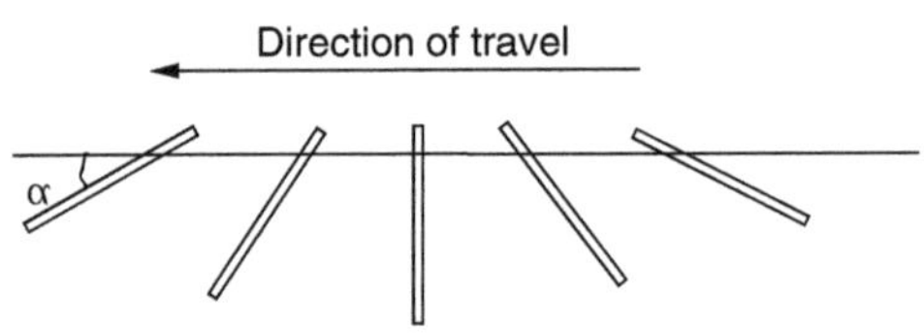

Fig. 1.1 Rake angle (α) is the angle between the tine and the horizontal in the direction of travel

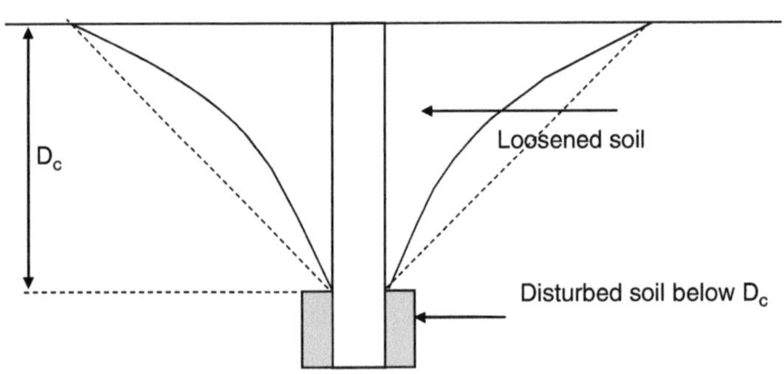

Fig. 1.2 Cross-section of soil worked by a tine. D_c corresponds to the critical depth, the maximum depth to which soil breaks up due to shear failure (Kostritsyn 1956). The dotted line corresponds to break-up at 45° from the side of the tine at D_c

1.3 Model Development

Models for predicting draught requirement are generally developed from soil mechanics within civil engineering. The most common models are based on calculation of the forces required for shear failure in front of a retaining wall (McKyes 1989). In this case the failure zone is very wide relative to its depth. This corresponds to soil break-up by a wide blade and is called two-dimensional soil cutting, since the end effects of the blade are neglected. Compressive forces are built up in front of the blade and the soil is assumed to fail according to the Mohr–Coloumb failure criterion. As the implement advances, it produces a succession of failure surfaces (Fig. 1.3a). In Fig. 1.3a, the shear plane is drawn for a frictionless tine. In reality, there is friction and adhesion between the soil and the tine, which makes the shape of the shear plane more complicated (Fig. 1.3b). The first part (i.e. the part closest to the tine tip) of the failure line is often given the shape of a logarithmic spiral, and the second part is given as a straight line (Osman 1964).

The force acting on a wide blade as given by Reece (1965) is:

$$F = (\gamma z^2 N_\gamma + czN_c + c_a z N_a + qzN_q)w \tag{1.1}$$

where F is the force, γ is specific weight of soil, z is the working depth, c is the cohesion, c_a is the adhesion between tine and soil, q is the surcharge, N_γ, N_c, N_a and

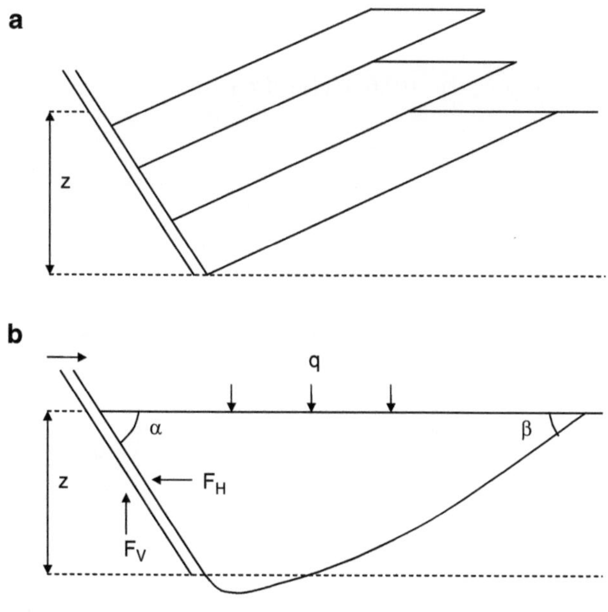

Fig. 1.3 (**a**) Creation of successive failure surfaces during tillage. (**b**) Shear failure in front of a tine, initial failure. After Stafford (1984)

N_q are dimensionless constants and w is the width of the blade. The N_x factors are functions of the angle of internal friction, the angle of soil–metal friction and the rake angle, and the four terms correspond to gravitational, cohesive, adhesive and surcharge effects. Numerical solutions for the dimensionless factors have been developed by Sokolovski (1965), Hettiarachi et al. (1966) and Hettiarachi and Reece (1974) to cover a range of values for angle of internal friction, angle of soil–metal friction and rake angle. Soil–metal adhesion is considered to have a small effect (Stafford and Tanner 1983b). Hettiarachi and Reece (1974) developed Eq. 1.1 by combining the cohesive and adhesive terms.

Hettiarachi et al. (1966) used Eq. 1.1 for the two-dimensional case. However, tillage in agriculture is most often carried out using narrow tines (width/depth ratio of 1 or less), for which the end effects of the tine (i.e. the soil disturbance on either side of the tine) cannot be neglected. Based on Eq. 1.1, semi-empirical models have also been developed for narrow tines, i.e. three-dimensional soil cutting (e.g. Hettiarachi et al. 1967; Godwin and Spoor 1977; McKyes and Ali 1977; Swick and Perumpral 1988; Kuczewski and Piotrowska 1998), using the shape of the crescent normally observed in front of narrow tines (Fig. 1.4). Wheeler and Godwin (1996) developed the model by Godwin and Spoor (1977) to also include the effect of implement velocity. The horizontal (= draught force) and vertical component of the soil cutting force can then be calculated as:

$$H = [(\gamma z^2 N_\gamma + czN_c + qzN_q)(w + z(m - 1/3(m - 1))) + \gamma v^2 N_a z(w + 0.6z)/g] \sin(\alpha + \delta) \tag{1.2}$$

$$V = -[(\gamma z^2 N_\gamma + czN_c + qzN_q)\,(w + z(m - 1/3(m - 1))) + \gamma v^2 N_a z(w + 0.6z)/g] \cos(\alpha + \delta) \tag{1.3}$$

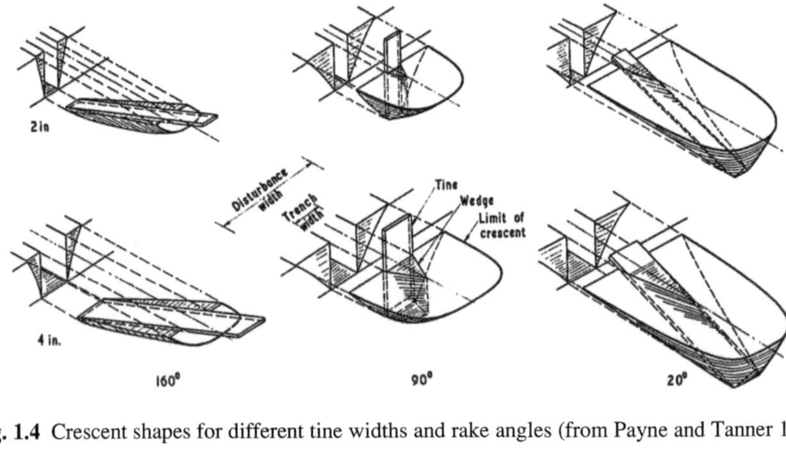

Fig. 1.4 Crescent shapes for different tine widths and rake angles (from Payne and Tanner 1959)

where H is the horizontal force, V is the vertical force, m is the rupture distance ratio (the ratio between forward rupture distance and working depth), N_a is a dimensionless factor for inertial effects, v is the velocity, g is acceleration due to gravity, α is the rake angle and δ is the angle of soil–metal friction, other notations as in Eq. 1.1. The term $z(m - 1/3(m - 1))$ corresponds to the side-effects of the tine, and the term $w + 0.6z$ corresponds to the effective area worked by the tine (Fig. 1.2). The model is available as a spreadsheet (Godwin and O'Dogherty 2007). In this chapter it is used to calculate examples of draught for a single tine (Fig. 1.5). Eqs. 1.2 and 1.3 are only for the case of shear failure, when the tine is working above the critical depth (Fig. 1.2). Godwin and Spoor (1977) also developed a model for determining the value of the critical depth and the draught of tines working below critical depth. Working below this depth causes lateral failure and

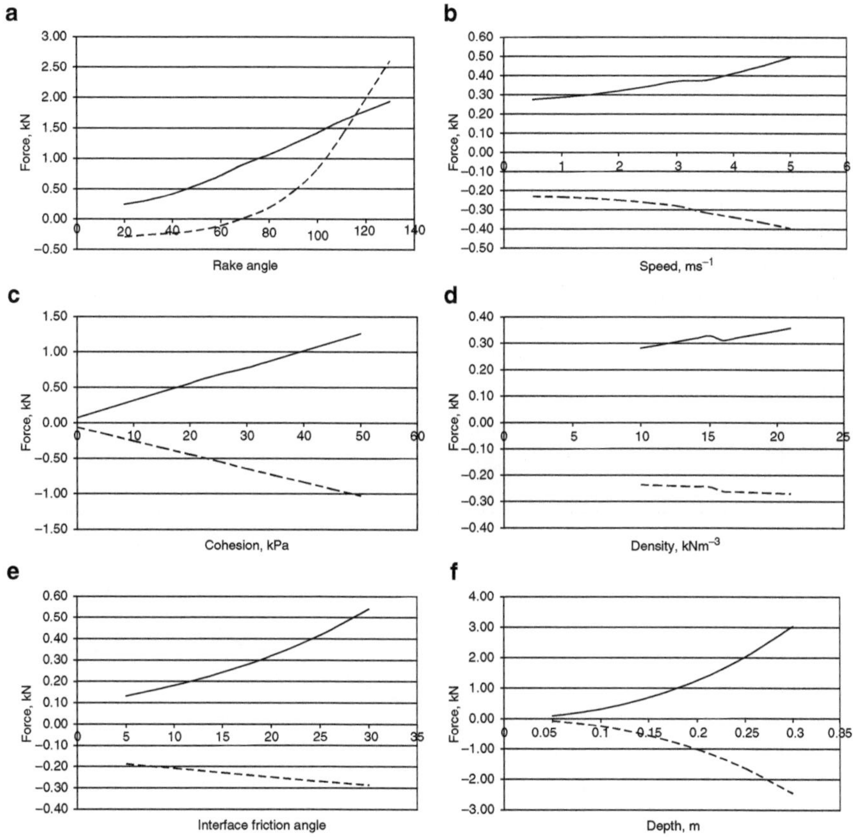

Fig. 1.5 Horizontal (*solid line*) and vertical (*dotted line*) force for a tine. Standard values: 30° rake angle, 0.05 m tine width, speed 2 m s^{-1}, 15 kN m^{-3} soil density, 20° soil-metal friction, 0.1 m working depth. The value of one parameter is varied in each graph, effects are shown for (**a**) rake angle, (**b**) speed, (**c**) shear strength, (**d**) soil density, (**e**) interface friction angle, (**f**) working depth. Calculations were made with the model and spreadsheet presented by Godwin and O'Dogherty (2007)

is not desirable from an agronomic point of view; therefore draught for working below this depth is not further discussed here.

The focus in this chapter is on the draught requirement of tined implements. Models to predict draught for other implements, such as mouldboard ploughs or disc implements, are also available (Godwin et al. 1985; Godwin and O'Dogherty 2007).

1.4 Effects of Rake Angle

The effects of rake angle on draught requirement have been widely studied (Payne and Tanner 1959; Dransfield et al. 1964; Stafford 1979; Makanga et al. 1996; Onwualu and Watts 1998; Aluko and Seig 2000). Fundamental work on the break-up of soil in front of tines was reported by Payne and Tanner (1959). The form of the crescent for different rake angles during shear failure is shown in Fig. 1.4. Payne and Tanner (1959) also showed the presence of a wedge on the face of the tine. This wedge is the "effective" tillage tool which exerts forces on the surrounding soil, while it is slowly rising along the tine. For small rake angles, there is a smaller horizontal distance between the front of the crescent and the tip of the tine. The width of the crescent is generally larger for small rake angles, and decreases greatly for very large rake angles. Increasing the rake angle increases the compressive forces exerted on the soil, this may increase fragmentation and create a finer seedbed compared with small rake angles. Compressive forces are especially large for backward rakes (rake angle $> 90°$), which may be efficient in

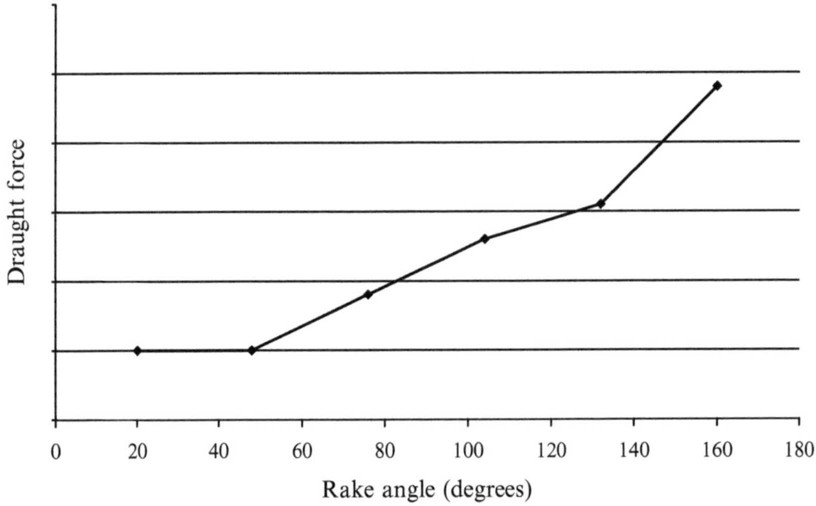

Fig. 1.6 Draught force for different rake angles expressed as the mean for a number of agricultural soils. After Payne and Tanner (1959)

breaking clods. Draught requirement is lower for small rake angles than for large (Fig. 1.6), as reported for example by Payne and Tanner (1959). They found the lowest draught requirement at 20° rake angle, increasing rapidly at angles greater than 45°; the energy per width of disturbed soil was approximately 8 times greater at 160° compared with 20° rake angle. Similar results have been reported by other authors (Dransfield et al. 1964; Aluko and Seig 2000). Model calculations also predict a large effect of rake angle, as shown in Fig. 1.5a. For the above reasons, small rake angles are most often used in primary cultivation and subsoiling, while more vertical or even backward tines (negative rake angle) are used in seedbed preparation.

As stated previously, shear failure during tillage has received most attention from researchers; however, tensile or brittle failure is also an important failure mechanism and is sometimes desirable from an agronomic perspective. Soil break-up by crack formation due to tensile forces in front of the tillage tool may reduce draught compared with shear failure (Stafford and Geikie 1987; Hettiaratchi 1988). Aluko and Seig (2000) studied this point when soil failure during the action of tines changed from shear to tensile and they concluded that tensile failure was more likely to occur for small rake angles and high soil strength. In the three soils they investigated, the transition from tensile to shear failure occurred at 32°, 48° and 62° rake angles. This means that for small rake angles that gave the lowest draft requirement, soil failure was due to tensile failure in their investigation; however, tensile failure resulted in larger soil clods compared with shear failure.

Different rake angles result in different directions (upward or downward) of the horizontal force and hence the resultant force during tillage. The adhesion and friction between the tine and the soil create an upward force, which means that for a vertical tine, the vertical force is directed upwards. In calculations using Eq. 1.3, there is no net vertical force at an angle of $90° - \delta$, where δ is the soil–metal friction angle. For a typical value of δ of approximately 22.5° this occurs at 67.5° rake angle (Godwin 2007; Fig. 1.5a). A similar value has been found in field measurements (e.g. Dechau and Yusu 1992; Aluko and Seig 2000). However, Payne and Tanner (1959) found no net vertical force at approximately 45° rake angle in three agricultural soils. Aluko and Seig (2000) found that for a rough metal surface, the change in direction of vertical force occurred at smaller rake angles than for a polished surface, in accordance with model predictions.

1.5 Effect of Implement Speed

Increasing speed means that a tine moves faster through the soil, and that shear failure in front of the tine occurs more frequently. The force, as calculated by Eq. 1.1, is the same irrespective of speed; however, in general there is a speed effect, with the draught force increasing with speed. This is often attributed to the acceleration of soil, which makes it natural to insert a factor proportional to the square of speed, as in Eq. 1.2. Olson and Weber (1966) discussed other possible explanations, such as increased shear strength and increased length of the failure path. In their results, there was a large increase in force with an increase in speed, especially for a tool with 90° rake angle compared with 30°. They concluded that acceleration of the soil, the length of the failure path, or the angle of inclination of the failure plane, could not explain the effects of speed. The effect of speed on the stress-strain curve and hence soil strength was a more probable explanation. Payne (1956) also found that increased acceleration of soil particles at higher speed only has a minor influence on draught. Schuring and Emori (1964) calculated the acceleration forces to be insignificant at speeds below $(5gw)^{½}$, where g is the gravitational acceleration and w the tine width. Wheeler and Godwin (1996) found that this critical limit could be increased to $[5g(w + 0.6d)]^{½}$, where d is the working depth. For a narrow tine with a width of 30 mm and a working depth of 250 mm, these two limits correspond to speeds of 4.4 and 10.7 km h^{-1}, respectively (Wheeler and Godwin 1996). Glancey et al. (1995) found speed effects for mouldboard ploughing below 7.2 km h^{-1} and concluded that the square of speed is normally a negligible factor.

There is also an interaction between speed and soil conditions. Dransfield et al. (1964) found that the maximum force was hardly affected by speed in loose soil, while it increased with speed in compacted, cohesive soil. Stafford (1979) also reported an increase in draught force with speed in cohesive soil. In plastic soil, with deformation due to plastic flow, the forces follow a decaying exponential form with speed (wet soil in Fig. 1.7). In the same soil at low water content, there was an increasing effect of speed on draught (Fig. 1.7). Stafford (1979) concluded that increasing draught with speed was due to increasing shear strength with increasing strain rate. He later developed two different relationships for draught versus speed

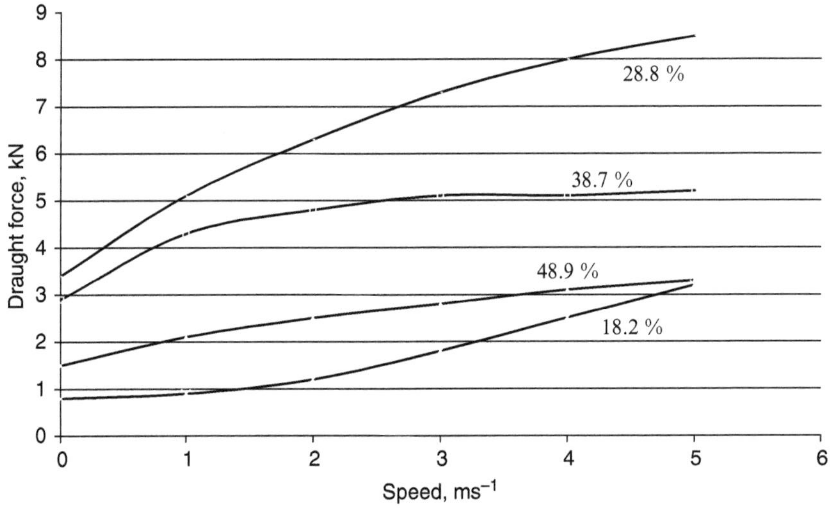

Fig. 1.7 Draught force for a tine as a function of speed in a clay soil. Different lines indicate different soil water contents. After Stafford (1979)

(Stafford 1984), one for brittle failure, similar to the speed term in Eq. 1.2, and one for plastic flow failure, with an exponential relationship. Reaves and Schafer (1974) examined draught versus speed relationships for a mouldboard plough and found draught to be equal to the square of speed for sand, while on clay soils the rate of increase in draught decreased at higher speeds. Thus for clay soils, acceleration was not an important factor for increase in draught.

1.6 Draught Requirement and Soil Strength

In Eq. 1.1, the factors determining soil properties are cohesion, soil internal friction and soil density. Stafford (1984) used only the second term in Eq. 1.1, since the cohesion term accounts for most of the draught force. Payne (1956) concluded that the effect of soil density is of minor importance due to the fact that the range of soil density in agricultural soils is much smaller than the range of cohesion. This is also shown in the model calculations in Fig. 1.5, where draught is a more or less linear function of the cohesion. While a large number of systematic studies have been carried out on the effect of rake angle or speed on draught, there are much fewer on the effect of soil strength.

In most studies, cohesion has been determined in triaxial tests, but this procedure is complicated and the equipment is expensive and often unavailable. Therefore, attempts have been made to correlate draught requirement to other soil strength parameters, especially soil penetration resistance (cone index) measured by a penetrometer (e.g. Eradat Oskoui et al. 1982; Eradat Oskoui and Witney 1982; Bowers 1989; Desbiolles et al. 1999; Arvidsson et al. 2004; Sahu and Raheman 2006). This is primarily because penetrometers are often available, and their measurements are relatively easy, fast and inexpensive. Penetration resistance can be seen as a composite soil property, governed by more basic properties including soil cohesion, soil compressibility and soil/metal friction (Dexter et al. 2007), i.e. properties also included in models to calculate draught during tillage (1.1–1.3).

Eradat Oskoui and Witney (1982) used the following equation to predict draught for a mouldboard plough from cone index data:

$$Z = K_1\mathrm{CI} + K_2\gamma v^2(1 - \cos\theta)/g \tag{1.4}$$

where Z is specific draught (kN m^{-2}), K_1 and K_2 are empirical coefficients, CI is cone index, v is plough speed, g is acceleration due to gravity, θ is mouldboard tail angle and γ is soil specific weight.

On the other hand, the relationship between draught and penetration resistance has been shown to be poor in some studies (Bowers 1989; Arvidsson et al. 2004; Fig. 1.8). Mulqueen et al. (1977) and Rahim et al. (2004) found a poor correlation between cohesion and penetration resistance, raising doubts as to whether the penetrometer is a useful tool for predicting draught during tillage.

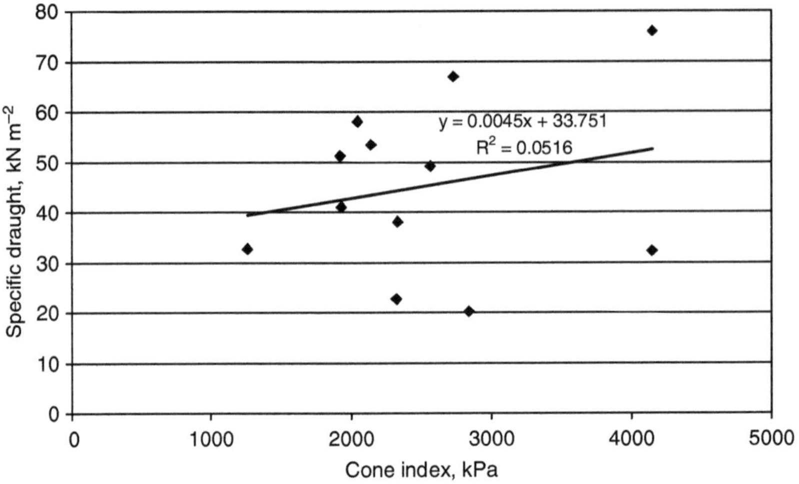

Fig. 1.8 Specific draught during mouldboard ploughing as a function of cone index (soil penetration resistance). Data from Bowers (1989)

Different methods for determining cohesion can also give different results. As mentioned previously, rate of shearing can increase with soil strength, which might explain speed effects on draught. This was found by Stafford and Tanner (1983a) but was not confirmed in measurements by Swick and Perumpral (1988). Stafford and Tanner (1982) compared different methods to determine cohesion and found much lower values for triaxial tests than for shear box, shear annulus and shear vane tests, with, for example, values measured by shear vane up to three times higher than those determined in a triaxial test. Schjønning (1991) also found much higher values of cohesion when measured by shear annulus compared with shear box. Thus, there is a need for a method to determine soil strength that is easy to use yet accurate in predicting draught force during tillage.

1.7 Draught and Soil Fragmentation

In most studies on draught during tillage, the value has been determined in relation to working depth or the area of soil worked. To determine the efficiency of tillage, the energy input should also be related to the results in terms of soil fragmentation, as can be measured by the aggregate size distribution (see Chap. 7). Since fragmentation of soil includes breaking bonds between soil particles and aggregates, it is reasonable to quantify the effectiveness in terms of energy in relation to the surface area of aggregates produced by tillage (Berntsen and Berre 1993). The surface area can be determined by sieving the soil into classes with different particle diameter (Hadas and Wolf 1983). The increase in area per unit mass can then be calculated according to Berntsen and Berre (1993) as:

$$\Delta A = \frac{6}{\gamma}\left(\frac{1}{D_2} - \frac{1}{D_1}\right) \tag{1.5}$$

where ΔA is the change in soil surface area, γ is the soil specific weight, D_2 is the aggregate mean diameter before tillage and D_1 is the aggregate mean diameter after tillage.

Berntsen and Berre (1993) then related this change in area to the draught force F in order to produce a factor f, which describes the efficiency in soil fragmentation:

$$f = \frac{6}{\gamma F}\left(\frac{1}{D_2} - \frac{1}{D_1}\right) \tag{1.6}$$

The same concept was used by Arvidsson et al. (2004), who found a higher efficiency in soil fragmentation with a disc implement compared to a mouldboard plough or a tined implement. However, the energy requirement for soil fragmentation has been studied much less than the draught requirement for soil break-up.

1.8 Ways of Reducing Draught

Tillage implements should be designed to have small energy requirements in relation to the desired outcome of tillage. As can be concluded from Figs. 1.5 and 1.6, implements for break-up of large soil volumes should generally have small rake angles.

Spoor and Godwin (1978) found that adding wings to a tine increased the soil volume worked much more than it increased the draught force, thereby substantially reducing the specific draught. Working the soil in two steps, with a gradual increase in working depth, can reduce draught for tined implements compared with one pass to the desired depth (McKyes 1989). Godwin et al. (1984) found that shallow tines, working ahead of deeper tines at intermediate positions, gave a small increase in draught but a large increase in disturbed area, thereby reducing specific draught. Similarly, Spoor and Godwin (1978) found that putting chisel tines before a subsoiler decreased specific draught. Working the soil in steps is especially important if it can avoid working below the critical depth, which is very unfavourable in terms of specific draught.

The angle of soil–metal friction has a large influence on draught in model calculations (Fig. 1.5e). However, Aluko and Seig (2000) did not find any great effect of metal surface roughness on the magnitude of the horizontal (draught) force, but mainly in the direction of the resulting force. The soil–metal friction angle is also affected by soil type and soil water content (Yusu and Dechau 1990), and may increase with increasing speed (Stafford and Tanner 1983b). Furthermore, if soil builds up on the tine, friction occurs between soil particles and between soil and metal. The interface friction angle may then be assumed to be equal to the soil internal friction angle (Wheeler and Godwin 1996).

The sharpness and thickness of the point of tines or plough bodies may also be important. Fielke (1996) found a large increase in draught (up to 80%) and an increased upward vertical force for a blunt cutting edge compared with a sharp edge. Vibrating tines may reduce draught compared with rigid tines (Berntsen et al. 2006). The reason for this reduction in draught for a flexible tine is not clear, but may be due to a lowering of speed before maximum force is exerted and the soil fails (Berntsen et al. 2006).

One of the most important factors determining draught is generally the tillage depth, as can be seen in model calculations on the draught of a tine (Fig. 1.5f). Although the draught requirement of a tine increases dramatically with an increase in depth, this may not cause a large increase in specific draught, since the loosened area increases more than linearly with depth. Assuming a break-up angle of 45° at the side of the tine, the area worked amounts to $w \times d + 2 \times d/2 = d(w + d)$. Willatt and Willis (1965) examined the trough made by a tine and found the area to be close to this value, while Wheeler and Godwin (1996) found it to be $d(w + 0.6d)$. Arvidsson et al. (2004) found an increase in specific draught with working depth for a chisel plough, but not for a mouldboard plough, while Owende and Ward (1996) also found no clear correlation between tillage depth and specific draught for mouldboard ploughing. Unfortunately, in most measurements of soil draught the area of soil worked is not determined, so specific draught cannot be calculated. In model calculations, specific draught increases with an increase in depth-to-width ratio of a tine (McKyes 1989). This was also found in measurements by Desir (1981), and is consistent with Arvidsson et al. (2004), who found a lower specific draught for a mouldboard plough compared with a chisel plough. Other comparisons have been made of draught requirements between mouldboard ploughs and tined implements but, as stated previously, the actual working depth is not measured carefully enough to allow comparisons in specific draught between these implements.

1.9 Conclusions

Models to predict draught requirement during tillage are generally based on equations to calculate forces on a retaining wall for shear failure of soil. Most publications show a good correlation between measured and predicted values using these models; however, most experiments are based on measurements in soil bins and do not consider the tillage outcome, e.g. the aggregate size distribution. Future research should place more emphasis on draught requirement under natural field conditions, where soil strength and structure may be very different from that of (remoulded) soil in soil bins. The energy requirement should also be defined in relation to the area of soil worked (specific draught) and, especially, the tillage outcome. There is also a need for a simple field method to determine soil strength in a way that is useful for predictions of draught requirement. Future studies on a combination of draught requirement and tillage outcome would be useful in implement design and in guiding farmers in their choice of implement.

References

Aluko OB, Seig DA (2000) An experimental investigation of the characteristics of and conditions for brittle failure in two-dimensional soil cutting. Soil Till Res 57:143–157

Arvidsson J, Keller T, Gustafsson K (2004) Specific draught for mouldboard plough, chisel plough and disc harrow at different water contents. Soil Till Res 79:221–232

Berntsen R, Berre B (1993) Fracturing of soil clods and the soil crumbling effectiveness of draught tillage implements. Soil Till Res 28:79–94

Berntsen R, Berre B, Torp T, Aasen H (2006) Tine forces established by a two-way model and the draught requirement of rigid and flexible tines. Soil Till Res 90:230–241

Bowers CG (1989) Tillage draft and energy measurements for twelve southeastern soil series. ASAE paper No. 86-1524.

Dechau Z, Yusu Y (1992) A dynamic model for soil cutting by blade and tine. J Terramechanics 29:317–327

Desbiolles JMA, Godwin RJ, Kilgour J, Blackmore BS (1999) Prediction of tillage implement draught using cone penetrometer data. J Agric Engng Res 73:65–76

Desir FL (1981) A field evaluation of the wedge approach to the analysis of soil cutting by narrow blades. M.Sc. Thesis, McGill Univ., Dept. Agric.Eng., Montreal, Canada.

Dexter AR, Czyz EA, Gate OP (2007) A method for prediction of soil penetration resistance. Soil Till Res 93:412–419

Dransfield P, Willatt ST, Willis AH (1964) Soil-to-implement reaction experienced with simple tines at various angles of attack. J Agric Engng Res 9:220–224

Eradat Oskoui K, Rackham DH, Witney BD (1982) The determination of plough draught-Part II. The measurement and prediction of plough draught for two mouldboard shapes in three soil series. J Terramechanics 19:153–164

Eradat Oskoui K, Witney BD (1982) The determination of plough draught-Part I. prediction from soil and meteorological data with cone index as the soil strength parameter. J Terramechanics 19:97–106

Fielke JM (1996) Interactions of the cutting edge of implements with the soil. J Agric Engng Res 63:61–72

Glancey JL, Shrini K, Upadhyaya SK (1995) An improved technique for agricultural implement draught analysis. Soil Till Res 35:175–182

Godwin RJ (2007) A review on the effect of implement geometry on soil failure and implement forces. Soil Till Res 97:331–340

Godwin RJ, O'Dogherty MJ (2007) Integrated soil tillage force prediction models. J Terramechanics 44:3–14

Godwin RJ, Seig DA, Allot M (1985) The development and evaluation of a force prediction model for agricultural discs. In: Proceedings of the International conference on Soil Dynamics, Auburn, AL, pp. 250–263.

Godwin RJ, Spoor G (1977) Soil failure with narrow tines. J Agric Engng Res 22:213–288

Godwin RJ, Spoor G, Soomro MS (1984) The effect of tine arrangement on soil forces and disturbance. J Agric Engng Res 39:47–56

Hadas A, Wolf D (1983) Energy efficiency in tilling dry clod-forming soils. Soil Till Res 3:47–59

Hettiarachi DRP, Reece AR (1974) The calculation of passive soil resistance. Geotechnique 24:289–310
Hettiarachi DRP, Witney BD, Reece AR (1966) The calculation of passive pressure in 2-D failure. J Agric Engng Res 11:89–107
Hettiarachi DRP, Witney BD, Reece AR (1967) Symmetrical three-dimensional soil failure. J Terramechanics 4:45–67
Hettiaratchi DRP (1988) Theoretical soil mechanics and implement design. Soil Till Res 11:325–347
Koolen AJ, Kuipers H (1983) Agricultural soil mechanics. Springer, Berlin
Kostritsyn AK (1956) Cutting of a cohesive soil mechanism with knives and cones. Translation 58. NIAE, Silsoe, Bedford
Kuczewski J, Piotrowska E (1998) An improved model for forces on narrow soil cutting tines. Soil Till Res 46:231–239
Makanga JT, Salokhe VM, Gee-Clough D (1996) Effect of tine rake angle and aspect ratio on soil failure patterns in dry loam soil. J Terramechanics 33:233–252
McKyes E (1989) Agricultural engineering soil mechanics. Elsevier, New York
McKyes E, Ali OS (1977) The cutting of soil by narrow blades. J Terramechanics 14(2):43–58
Mulqueen J, Stafford JV, Tanner DW (1977) Evaluation of penetrometers for measuring soil strength. J Terramechanics 14:137–151
Olson DJ, Weber JA (1966) Effect of speed on soil failure patterns in front of model tillage tools. Trans Soc Automot Eng 74:298–310
Onwualu AP, Watts KC (1998) Draught and vertical forces obtained from dynamic soil cutting by plane tillage tools. Soil Till Res 48:239–253
Osman MS (1964) The mechanics of soil cutting blades. J Agric Engng Res 9:313–328
Owende PMO, Ward SM (1996) Characteristic loading of light mouldboard ploughs at slow speeds. J. Terramechanics 33:29–30
Payne PJC (1956) The relationship between the mechanical properties of soils and the performance of simple cultivation implements. J Agric Engng Res 1:23–50
Payne PJC, Tanner DW (1959) The relationship between rake angle and the performance of simple cultivation implements. J Agric Engng Res 4:312–325
Reaves CA, Schafer RL (1974) Force versus width of cut for mouldboard bottoms. A.S.A.E. Paper No. 74-1588.
Rahim A, Prasad SN, George KP (2004) Dynamic cone penetration resistance of soils–theory and evaluation. Geotechnical Special Publication (126 II), 1755–1766.
Reece AR (1965) The fundamental equation of earthmoving mechanics. Proc. Symp. Eart-moving Machinery. I. Mech. E. 179:8–14
Sahu RK, Raheman H (2006) Draught requirement of agricultural implements using reference tillage tools in sandy clay loam soil. Biosystems Eng 94:275–284
Schjønning P (1991) Soil mechanical properties of seven Danish soils. Tidskrift for planteavls Specialserie, Beretning nr S 2176-1991. Lantbruksministeriet, Statens Planteavlsforsøg, Danmark.
Schuring DJ, Emori RI (1964) Soil deforming processes and dimensional analysis. Paper 897c. SAE New York.
Sokolovski VV (1965) Statics of soil media. Pergamon Press, Oxford
Spoor G, Godwin RJ (1978) An experimental investigation into the loosening of soil by rigid tines. J Agric Engng Res 23:243–257
Stafford JV, Tanner DW (1983a) Effect of rate on soil shear strength and soil–metal friction I. Shear strength. Soil Till Res 3:245–260
Stafford JV, Tanner DW (1983b) Effect of rate on soil shear strength and soil–metal friction I. Soil–metal friction. Soil Tillage Res 3:321–330
Stafford JV, Tanner DW (1982) Field measurent of soil shear strength and a new design of field shear meter. Proceedings of the 9th conference of ISTRO, Osijek, Yogoslavia.

Stafford JV (1979) The performance of a rigid tine in relation to soil properties and speed. J Agric Engng Res 24:41–56
Stafford JV (1984) Force prediction models for brittle and flow failure of soil by draught tillage tools. J Agric Engng Res 29:51–60
Stafford JV, Geikie A (1987) An implement configuration to loosen soil by inducing tensile failure. Soil Till Res 9:363–376
Swick WC, Perumpral W (1988) A model for predicting soil-tool interaction. J Terramechanics 25:43–56
Wheeler PN, Godwin RJ (1996) Soil dynamics of single and multiple tines at speeds up to 20 km/h. J Agric Engng Res 63:243–250
Willatt ST, Willis AH (1965) A study of the trough formed by the passage of tines through soil. J Agric Engng Res 10:1–4
Yusu Y, Dechau Z (1990) Investigation of the relationship between soil–metal friction and sliding speed. J Terramechanics 27:283–290

Chapter 2
Influence of Soil Tillage on Soil Compaction

Barbora Badalíková

2.1 Definition of Soil

Soil is one of the necessary requirements for human existence and an essential component of human civilization as a whole. It is a fundamental prerequisite for agricultural production and is closely connected with food supply. Krejčíř (1990) defines soil as a heterogeneous, multi-phase living system which is characterised by certain physical, chemical and biological properties. The soil forms the Earth's immediate surface – the pedosphere – which is derived from the Earth's crust by weathering. Soil must be seen as being part of the whole complex, as a component of the natural environment which, along with the atmosphere, hydrosphere and biocoenosis, forms a functional ecological system called the ecosystem. Soil is the product of the soil-forming processes, in which the parent material changes into large soil groups and soil types. The components of the environment influence soil and soil has an effect on the other components of the environment as well. This interaction means that any intervention in any one of the components of the ecosystem is an intervention in the ecosystem as a whole (Prax et al. 1995).

2.2 Soil Fertility

The basic characteristic of soil that is assessed is soil fertility. It can be defined as the capacity of soil to produce cultivated agricultural crops. Soil fertility, however, cannot be characterised just by one or several of its properties, as it is the result of a large complex of interacting properties. Among the principal soil properties which determine soil fertility are its physical properties. These properties may be highly

B. Badalíková
Agricultural Research, Ltd, Zahradní 1, 664 41 Troubsko, Czech Republic
e-mail: badalikova@vupt.cz

A.P. Dedousis and T. Bartzanas (eds.), *Soil Engineering*, Soil Biology 20,
DOI 10.1007/978-3-642-03681-1_2, © Springer-Verlag Berlin Heidelberg 2010

variable (temperature, soil water content, etc.), resulting in qualitative changes to soil properties and consequently in changes to soil fertility. This is related to a sufficient amount of organic matter in soil (Pranagal 2004), which, as has been said, guarantees a balance between individual physical properties under different climatic conditions during vegetation.

The term soil fertility covers

(a) Potential or natural fertility, and
(b) Real or effective fertility.

The term potential or natural soil fertility refers to the soil which has never been used for agricultural production and has never been exploited by humans. It develops naturally from the surface material of the Earth's crust by weathering and from residues of organic and living organisms living in the soil and on the soil surface under the influence of the climate. These are natural, virgin soils.

Real or effective fertility is the product of human activities which aim to create conditions for high productive capacity of soil to the benefit of mankind. Humans have continuously cultivated the soil by applying various agronomic measures which cause changes not only to biological, chemical and physical properties of soil but also to the process of soil formation. The result may be positive cultivation which turns a less valuable type of soil into a more valuable one (change in water regime or saline soil remediation). On the other hand, negative cultivation may also occur. This means deterioration of soil fertility by reducing or worsening humus quality, deterioration of soil structure, nutrient depletion, reduction in microbial activity, or soil compaction.

Therefore, the nature of soil cultivation, i.e. human intervention in soil genesis and evolution, resides in the fact that, by changing quantitative and qualitative participation and relationships of factors, humans alter the environmental conditions making the original soil-forming processes more or less profound. As a result, the soil forming process acquires a new dimension and the soil new properties, either positive or negative.

Keeping an optimum balance between soil properties in the soil management system is essential in regulating and controlling basic conditions of soil fertility.

2.3 Soil Tillage

Soil tillage is closely related to the level of soil compaction. Soil tillage, as an agronomic practice that requires considerable expense and high-energy inputs, is to create favourable conditions for good stand establishment, stand growth and development and crop yields. One of the main goals of soil tillage is influencing soil processes, predominantly modification of the physical, chemical and biological properties of soil.

One of the basic soil properties affected by tillage is the bulk density, although soil tillage by itself has a minimal effect on its stabilisation. The bulk density is most frequently used as the most important quantity expressing the physical properties of soil. Its increase or reduction has an effect on the rate of mineralization of organic matter. Higher bulk density alters the ratio of water-to-air capacity proportionally in favour of water capacity. It reduces total soil porosity and increases the proportion of capillary pores that contribute, to some extent, to improved water regime and water supply to plants in the course of vegetation (Badalíková and Kňákal 2000).

Various soil tillage practices have an effect on the soil environment and subsequently on soil fertility, and farmers should manage the soil in such a way as to prevent soil damage and irreversible degradation processes.

In the agricultural practices of soil tillage and establishment of stands of major field crops, the agronomic operations and practices which can be characterised predominantly by reduction in tillage depth and lower tillage intensity, combining more field operations, including crop planting, and leaving crop residues on the soil surface or in the topsoil, have become quite popular (Tebrügge and Düring 1999). These soil tillage and planting systems are generally known as minimum tillage and soil erosion control practices.

Soil tillage predominantly changes the physical status of soil on which its water, air, biological and thermal regimes are heavily dependent. It is also a valuable tool for weed, pest and disease control and helps improve soil fertility and create optimum conditions for plant growth and development. Opinions on soil tillage differ but they are all related to site conditions. Soil tillage practices are dependent on soil and climatic conditions. The reason for introducing reduced tillage systems or eliminating some field operations was predominantly the necessity to use alternative soil tillage practices in places exposed to wind and water erosion, where there was not enough time needed for carrying out all cultivation operations in the right sequence or not enough farm machinery for these operations. In introducing these minimum tillage systems it is necessary to observe certain rules and to implement a system of agronomic practices, including good weed, disease and pest control. The occurrence of weeds is higher under these technologies and therefore, it is necessary to have a good knowledge of the efficacy and application of herbicides to avoid great expense in this area.

An important proven factor in minimum tillage technologies is maintaining soil fertility, i.e. maintaining good soil structure which is created by a mixture of soil aggregates of different size and shape, porosity, mechanical resistance and water stability. Soil structure determines the water and air regime and, as a result, the biological and nutrient regime of the soil.

Soil tillage systems considerably affect soil permeability. Soil infiltration is directly proportional to the stability of soil structure (Tisdall and Adem 1986), pore size, volume and structure (Patel and Singh 1981; Ankeny et al. 1990; Badalíková and Hrubý 2006). Long-term zero-tillage or conventional tillage, on the other hand, can change the volume of pores, aggregate stability and organic matter content and consequently the entire soil structure (Drees et al. 1994; Lal et al. 1994; Singh et al. 1994; Diaz-Zorita et al. 2004). This may also bring about changes to soil properties by influencing infiltration rate of soil and soil water movement.

2.4 Soil Compaction

Soil compaction is an important process of soil degradation affecting the crop-producing role of soil, such as its vulnerability to soil erosion, soil water and nutrient availability, and natural biological activity of soil, etc. Soil compaction is manifested by total deterioration of the regime of the physiological profile of soil. Soil compaction is caused by two factors:

- Natural – conditioned by genetic properties of soil, pressure of roots penetrating the soil, kinetic energy of rain, effects of water logging and successive frost-free winters
- Artificial – caused by humans
  - Direct – by field machinery – passes, pressure and drive slip of machines
  - Indirect – reducing soil strength to compaction by incorrect management practices such as insufficient supply of organic fertilisers, bad choice of fertilisers, mistakes in crop rotations, continuous growing of crops, etc.

From the agricultural point of view, compacted soil has low porosity, low water and air permeability, and increased requirements for traction power in seedbed preparation. It has been proved that, with total porosity of soil less than 45 vol.%, the conditions are not good for more stable forms of humus.

Soil compaction often occurs on wet sandy soil in which heavy machines make tracks as well as structural and textural changes in the vertical direction. In a wet soil with 20–40% of clay the wheel traffic makes a track in which the ratio of the part pushed aside to the part pushed down is about 1:3. The soil is less compacted but more worked up in depth and width. On a wet clay soil the track is formed without hardly any changes in bulk density. The soil is incompressible and is perfectly deformable; however, the clay soil becomes thoroughly moulded. Lhotský et al. (1984) determined that at a pressure of 0.15 MPa soil moisture of loam soils must not exceed 80% of field water capacity and 90% in sandy soils. Soil moisture content is directly connected to soil compaction. The biggest threat to soil structure (soil compaction) is the passes of agricultural machines at high soil moisture. It is stated that one pass of the tractor with a contact force of 0.15 PMa becomes evident at a depth to 0.3–0.4 m, and a repeated pass at a depth to 0.6 m (Lhotský et al. 1984).

A significant factor in the process of compaction itself, or loosening and determining the state of compaction, is water content of the soil especially when penetrometry is employed. With higher soil moisture content compressibility increases in the parabolic function to the apex which corresponds approximately to 80% of field water capacity. From this point compressibility decreases (Baver et al. 1972).

Penetrometric measurements are based on detecting the force necessary for pushing the standard steel cone to the soil. Its advantage is great rapidity and the possibility of interpreting results for the whole soil profile under study (Badalíková and Pokorný 2007).

Table 2.1 Limit values of soil properties

Soil property	Soil type					
	j[c]	jv[c], jh[c]	h[c]	Ph[c]	hp[c]	p[c]
Porosity (vol.%)	<48	<47	<45	<42	<40	<38
Reduced bulk density (g cm^{-3})	>1.35	>1.40	>1.45	>1.55	>1.60	>1.70
Penetrometric soil strength (MPa) –	2.8–3.2	3.2–3.7	3.7–4.2	4.5–5.0	5.5	6.0
at soil moisture content (wt.%)[a]	28–24	24–20	18–16	13–15	12	10
Minimum air capacity (vol.%) [b]	<10	<10	<10	<10	<10	<10

[a]If soil moisture lies outside the interval given in the following line, then for each weight percent of soil moisture either add 0.25 MPa to the value of critical resistance (*lower water content*) or subtract 0.25 MPa from the value of critical resistance (*higher water content*)

[b]10% is the average value of minimum air porosity; in vertical pore orientation the limit value reduces to 8 vol.%, in horizontal pore orientation it increases up to 15 vol.%. The limit value varies with crops (root crops 12%, cereals 10%, and forage crops 8%) (After Lhotský et al. 1984)

[c]j = c [clay]; jv, jh = ce, cl [cleyey, cleyloamy]; h = l [loamy]; ph = sl [sandy-loamy soil]; hp = ls [loamy-sandy soil]; p = s [sandy soil]

As for soil compaction measurements, an indirect correlation between penetrometric resistance of soil and soil moisture has almost always been confirmed. The relationship between soil moisture and soil compaction has also another aspect – soil compaction changes the quality of pores and, to some extent, increases the maximum capillary water capacity. As a result, the average soil moisture changes.

The limit values of soil properties for detrimental soil compaction were determined by Lhotský et al. (1984), see Table 2.1.

An almost linear dependence was proved between soil resistance measured by the penetrometric technique and the bulk density, confirming the practical applicability of the penetrometer. Rátonyi (1998) also discovered that long-term no-tillage practices or shallow disking had an effect on soil compaction and water content and that the relationship between the penetrometric resistance of soil, bulk density and soil water was linear.

Changes caused by soil compaction deteriorate soil physical properties and are most evident in the reduced bulk density which affects the whole complex of soil physical properties such as porosity, air and water capacity, soil thermal conductivity, etc. At the same time there are some changes in soil water content, availability and movement. Water is a very important factor, not only in biomass production of cultivated plants, but also in maintaining soil fertility from the physical and the chemical viewpoint. Both soil water surplus and deficit are detrimental.

It is evident that soil compaction significantly affects the behaviour and the rate of physical–chemical and biological processes. However, with proper management we can influence the reproductive process of cultivated plants (Hraško and Bedrna 1988). Physical properties of soil have an effect on soil microclimate of vegetation, especially on water, thermal and air regime. Soil loosening operations, especially ploughing, bring about considerable changes to reduced bulk density of soil. After loosening, the bulk density reduces and usually recovers in about 12 months (returns to its initial, natural state; Badalíková and Kňákal 1997). This ability is typical of good soils, especially chernozem. It was also discovered that a certain level of soil compaction is not always detrimental, it can have positive effects. It helps retain a higher content of soil water for a longer period, which is of vital importance especially in lower precipitation areas (Badalíková and Hruby 1998). It was proved that some crops, e.g. spring barley, respond positively to slight soil compaction. It is important to keep soil in good structural condition and not to adversely affect the formation of structural elements.

The main factors in eliminating the causes of soil compaction are soil tillage, field machinery and work management.

2.5 Monitoring of Soil Compaction – Experimental Results

Agronomic trials depend on assessment of physical and chemical parameters in the course of their observations.

The penetrometric technique is being tested to monitor soil compaction. This method is much faster and the results can be easily interpreted but, unlike the Kopecky method, it does not provide the whole range of results. Its accuracy is dependent on soil moisture which must be taken into consideration when the results are being interpreted.

Various soil tillage practices and soil compaction were studied in three localities of a sugar beet growing region with different soil and climatic characteristics (Badalíková and Pokorný 2007). Two of the localities were on chernozem soils and one on brown soil. Changes in soil compaction were measured by the penetrometer at depths to 0.45 m. Three soil tillage treatments were compared: treatment 1 – shallow loosening; treatment 2 – ploughing; treatment 3 – deep loosening. Each experimental treatment, with five replication measurements, was performed five times. Penetrograms (Badalikova and Pokorny 2007), which are graphical representations of the penetration resistance of soil (Figs. 2.1–2.4), were interpreted using a contact template and the average results can be seen in the following graphs and tables.

The results of penetrometric measurements in these agronomic trials with sugar beet, showed that on good chernozem soil, deep loosening is not so important because the soil is capable of recovering naturally. Ploughing was sufficient in this instance and, in other years, shallow loosening would suffice. Significantly higher soil resistance was recorded in brown soil in autumn when there were no differences between soil tillage systems.

A more accurate method of assessing soil compaction is determination of physical properties using Kopecky push tubes This method is very labour-consuming, provides only small samples of soil, and the results are often highly variable and difficult to interpret correctly for statistical purposes.

This method was chosen for field trials to study soil conditions in a sugar beet growing region which used a crop rotation system of winter wheat after a wheat crop the preceding year. The soil was classified as loamy, moderately heavy, degraded chernozem developed on loess. The physical properties of soil were

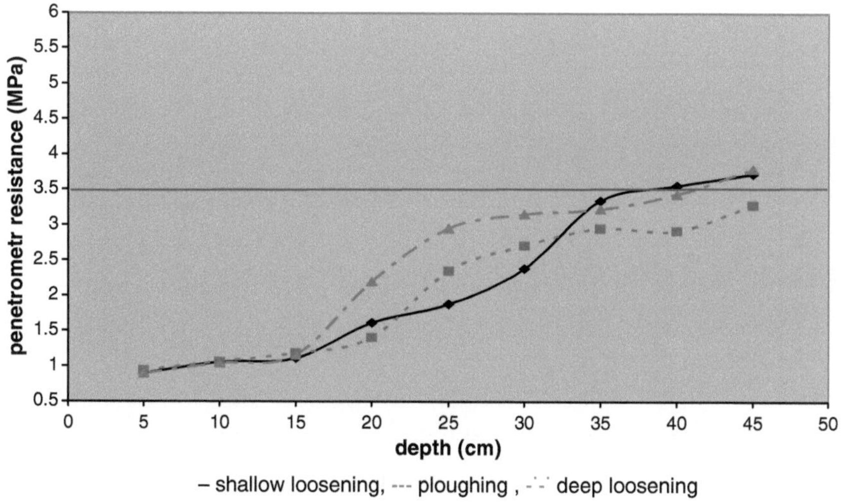

Fig. 2.1 Chernozem locality, spring measurement (Pokorný, 2007)

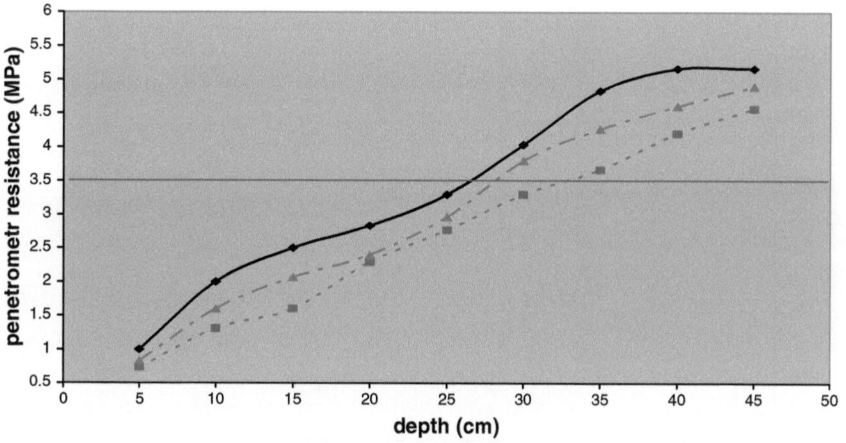

Fig. 2.2 Chernozem locality, autumn measurement (Pokorný, 2007)

examined under three different soil tillage treatments: treatment 1 – conventional tillage with ploughing; treatment 2 – minimum tillage with stubble incorporation; treatment 3 – no-till planting system.

It was found that the physical properties of soil were affected by different soil tillage operations used for winter wheat (Table 2.2). A higher bulk density and resulting lower porosity were recorded in minimum tillage and no-till planting systems. It was discovered that there is a relationship between soil compaction and soil water content, mostly in surface layers. In drier conditions during vegetation

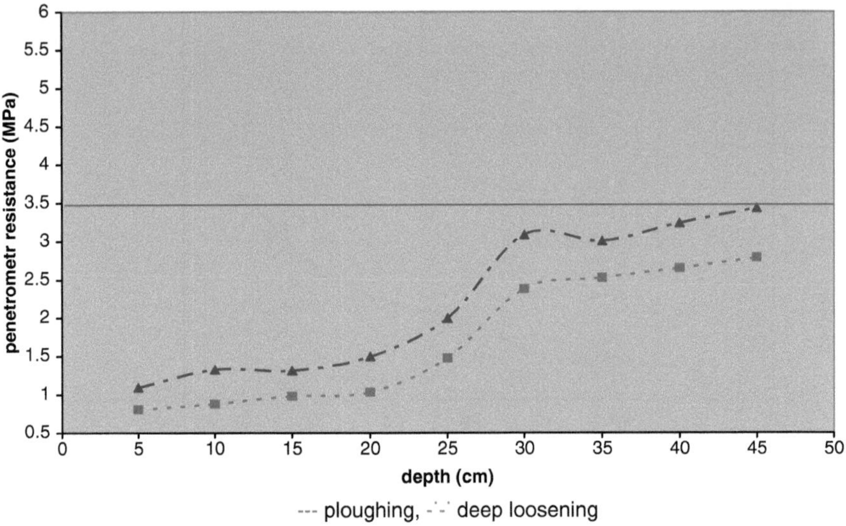

Fig. 2.3 Brown soil locality, spring measurement (Pokorný, 2007)

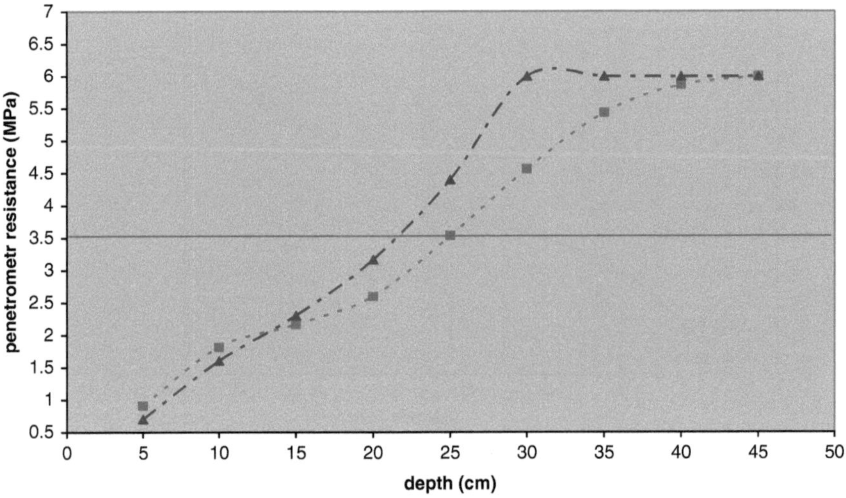

Fig. 2.4 Brown soil locality, autumn measurement (Pokorný, 2007)

there was water deficit in the treatment using ploughing, whereas treatments with stubble incorporation or no-till planting retained more soil moisture, due to better water-holding capacity.

The fact that water is retained for a longer period in compacted soil was also confirmed in other field trials in a maize growing region (warm and dry). These

Table 2.2 Soil physical properties in various soil tillage systems for winter wheat – sugar beet growing region

Soil tillage treatment	Depth (m)	Bulk density ($g \cdot cm^{-3}$)	Total porosity (vol.%)	Momentary content of		Maximum capillary capacity (vol.%)	Minimum air capacity (vol.%)
				Water (vol.%)	Air (vol.%)		
1	0.0–0.10	1.29	48.5	26.6	21.9	38.7	9.8
	0.10–0.20	1.55	38.1	26.0	12.1	36.2	2.0
	0.20–0.30	1.52	39.0	22.3	16.7	35.7	3.4
	Average	1.45	41.9	25.0	16.9	36.9	5.0
2	0.0–0.10	1.37	45.4	27.1	18.3	40.2	5.2
	0.10–0.20	1.66	33.4	30.0	3.4	29.5	3.9
	0.20–0.30	1.60	36.2	23.0	13.2	33.0	3.2
	Average	1.54	38.3	26.7	11.6	34.2	4.1
3	0.0–0.10	1.48	40.8	29.6	11.2	37.5	3.3
	0.10–0.20	1.63	34.9	25.9	9.0	29.6	5.3
	0.20–0.30	1.64	34.3	19.8	14.5	28.7	5.7
	Average	1.58	36.7	25.1	11.6	31.9	4.7

trials were carried out under field conditions on soils which were classified as moderately heavy, clayey loam, degraded chernozem developed on loess.

The soil environment was studied in a five-field crop rotation of winter wheat and maize under three soil tillage and planting systems: treatment 1 – medium ploughing, treatment 2 – minimum tillage, treatment 3 – no-till planting system.

A reduction in soil tillage intensity increased soil bulk density and decreased total porosity, whereas the maximum capillary capacity was not significantly affected by soil tillage and the total reduction in porosity means reductions in non-capillary pores. In the soil sampling procedures higher water content was recorded in treatment 2 (minimum tillage) and treatment 3 (no-till planting). This again confirmed that higher bulk density has a positive effect on soil water retention in both crops (Figs. 2.5 and 2.6).

Soil management resulting in soil compaction was also reported in a field trial in a potato growing region. In this instance, the negative effect of ploughing and soil tillage at an inappropriate time when soil moisture content was high was evident and had caused soil compaction, i.e. increase in the bulk density and reduction in porosity in different soil tillage systems (1 – ploughing, 2 – minimum tillage), which had negatively affected plant growth and development and aggravated conditions for subsequent soil tillage. The soil was loamy and the soil type was model brown soil.

The analysis of physical properties on this site revealed a high degree of degradation processes due to anthropogenic activities. Bulk density values given in Table 2.3 suggest deep soil compaction especially in the subsoil.

Penetrometric measurements were made on this site and high penetration resistance was found in deeper layers of the soil, corresponding to high values of reduced bulk density.

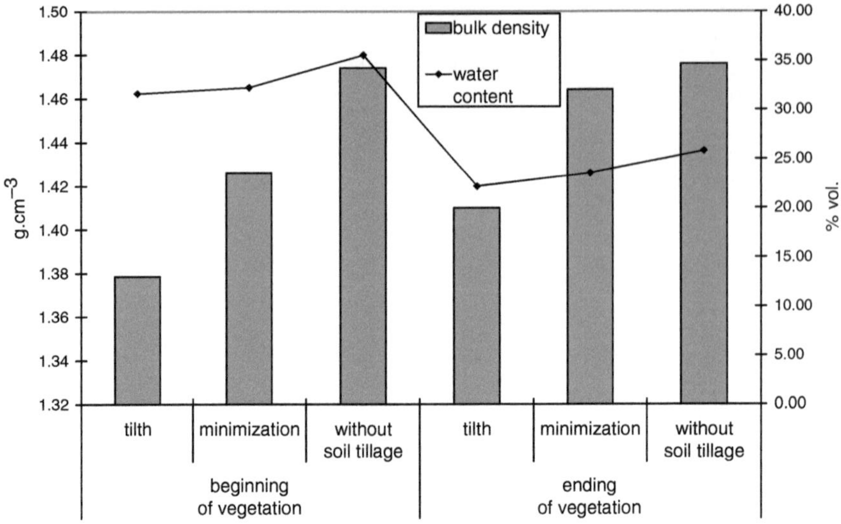

Fig. 2.5 Relationship between soil bulk density and momentary water volume under different soil tillage systems for winter wheat

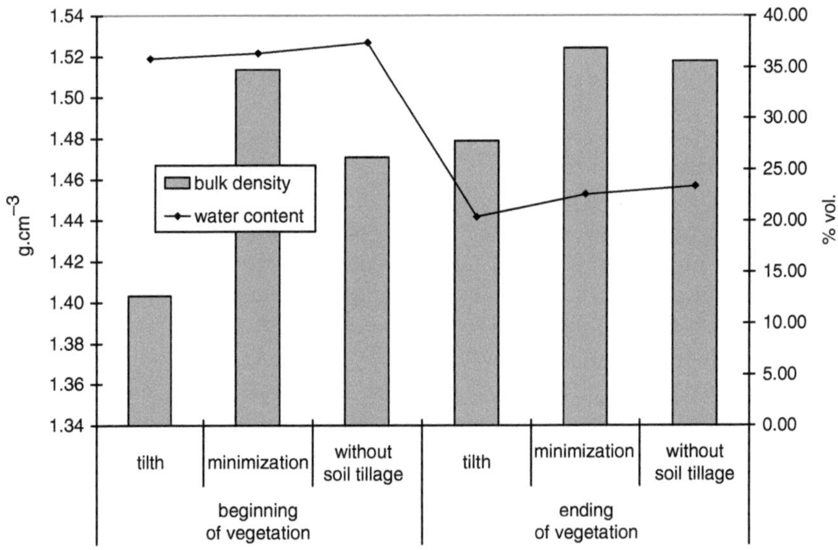

Fig. 2.6 Relationship between soil bulk density and momentary water volume under different soil tillage systems for maize

Table 2.3 Soil physical properties – potato growing region

Soil tillage treatments	Depth (m)	Bulk density ($g \cdot cm^{-3}$)	Total porosity (vol.%)	Momentary content of		Maximum capillary capacity (vol.%)	Minimum air capacity (vol.%)
				Water (vol.%)	Air (vol.%)		
1	0.0–0.10	1.49	44.14	17.75	26.39	37.24	6.90
	0.10–0.20	1.62	39.29	25.50	13.78	35.37	3.91
	0.20–0.30	1.69	37.03	19.92	17.11	33.32	3.71
	Average	1.60	40.15	21.06	19.10	35.31	4.84
2	0.0–0.10	1.54	42.04	17.17	24.87	34.86	7.18
	0.10–0.20	1.73	34.93	21.87	13.06	30.54	4.39
	0.20–0.30	1.75	34.74	20.79	13.95	31.77	2.97
	Average	1.67	37.24	19.94	17.29	32.39	4.85

Blecharczyk et al. (2007) found higher bulk density in the soil surface layer of direct sowings than in soils of other tillage systems. In the layer 10–20 cm, the bulk density of soil was higher in both direct sowings and surface tillage systems than in conventional tillage.

Dzienia et al. (2001) showed an increase in available potassium accumulation in soils of both direct sowing and reduced tillage systems compared to the conventional system.

2.6 Conclusion

Some types of soil degradation can be quite easily eliminated or at least their negative impact may be alleviated, usually by just obeying the rules of management for a particular soil. The care for soil physical properties and soil structure is very important if we want to retain soil plasticity and its quality for sustainable soil fertility. The quality of soil is the product of soil resistance to changes and the impact of degradation processes. From the agricultural point of view, the most important quality for maintaining soil stability is soil physical status which is mainly influenced by soil structure, porosity and compaction. Impairment of these qualities, mainly by human intervention, causes physical degradation of soil. Field machines may have a negative effect on soil environment but it depends largely on the external and internal soil environment. Passes of heavy field machines greatly affect soil properties in all their physical parameters. When the soil is wet, the first pass of the machine causes up to 90% compaction. Lower intensity of soil tillage increases reduced bulk density, reduces total porosity (at the cost of non-capillary pores) and reduces minimum air capacity, which directly correlates with soil compaction. However, this need not be negative; it depends on the type of soil, the amount of humus in the soil and the regenerative powers of the soil as to whether the soil is able to eliminate compaction and recover.

Acknowledgment This study is part of the research project no. QH72039 and no. 1G46038 that is financed by the grant of the National Agency for Agriculture Research of the Czech Republic Ministry of Agriculture.

References

Ankeny MD, Kaspar TC, Horton R (1990) Characterization of tillage and traffic effect on unconfined infiltration measurements. Soil Sci Soc Am J 54:s837–s840

Badalíková B, Hrubý J (1998) Physical soil properties at different systems of sugar beet cultivation. In: Proc.: Soil condition and crop production. Gödöllő, Hungary, pp 83–85

Badalíková B, Hrubý J (2006) Influence of minimum soil tillage on development of soil structure. In: Soil management for sustainability. Adv Geoecology 38:430–435

Badalíková B, Kňákal Z (1997) An influence of the management upon the physical properties of soil. In: Proc.: Agrochemical and ecological aspects of soil tillage. Tom 1A/97, Puławy, Poland, pp 55–58

Badalíková B, Kňákal Z (2000) Effect of different tillage systems for winter wheat on physical properties of soil. In: CD ISTRO-Conference 2000, Tillage at the Threshold of the 21st Century: Looking Ahead, Texas, USA

Badalíková B, Pokorný E (2007) Změny obsahu humusu při různém zpracování a utuženosti půdy k cukrovce. (The changes of humus content by different soil tillage and soil compaction in sugar beet). Agro 4:23–25

Baver LD et al. (1972) Soil physics. 4 edn. Wiley, New York p 498

Blecharczyk A, Małecka i, Sierpowski J (2007) Wpływ wieloletniego oddziaływania systemów uprawy roli na fizyko-chemiczne właściwości gleby. Fragm Agron 1:7–13

Diaz-Zorita M, Grove HJ, Murdock L, Herberck J, Perfect E (2004) Soil structural disturbance effects on crop yields and soil properties in a no-till production system. Agron J 96:1651–1659

Drees T, Karathanasis AD, Wilding LB, Blevins RL (1994) Micromorphological characteristic of long-term no-tillage and conventionally tilled soils. Soil Sci Soc Am J 58:s508–s517

Dzienia S, Pużyński S, Wereszczaka J (2001) Impact of soil cultivation systems on chemical soil properties. EJPAU, Agronomy, 4(2)

Hraško J, Bedrna Z (1988) Aplikované pôdoznalectvo. Bratislava, Príroda 47s

Krejčíř J (1990) Obecná produkce rostlinná. Skriptum, MZLU, Brno 133s

Lal R, Mahboubi AA, Fausey NR (1994) Long term tillage and rotation effects on properties of a central Ohio soil. Soil Sci Soc Am J 58:s517–s522

Lhotský J, Váchal J, Ehrlich J (1984) Soustava opatření k zúrodňování zhutnělých půd. Metodika pro zavádění výsledků výzkumu do zemědělské praxe č 14:38 s

Patel MS, Singh NT (1981) Changes in bulk density and water intake rate of a coarse textured soil in relation to different levels of compaction. J Indian Soc Soil Sci 29:s110–s112

Pranagal J (2004) Wpływ systemu uprawy na zawartość węgla organicznego w glebie. Annales UMCS Sec. E 59(1):1–10

Prax A, et al. (1995) Půdoznalství. Skriptum, MZLU, Brno, p 61s

Rátonyi T (1998) Evaluation of tillage effects on penetration resistance of chernozem soil. In: Proceedings, Soil condition and crop production. Gödöllő, Hungary, pp 100–103

Singh B, Chanasyk D S, Mcgill WB, Nyborg MPK (1994) Residue and tillage management effects on soil properties of typic Cryoboroll under continuous barley. Soil and Tillage Res 32s:117–133

Tebrügge F, Düring RA (1999) Reducing tillage intensity – a review of results from a long term study in Germany. Soil Till Res 53:15–28

Tisdall JM, Adem HH (1986) Effect of water content of soil and tillage on size-distribution of aggregates and infiltration. Aust J Exp Agric 26(2):193–195

Chapter 3
Vibrating Tillage Tools

László Fenyvesi and Zoltán Hudoba

3.1 Introduction

3.1.1 Draught Force Appearing in Soil Tillage

Draught force plays a key role when developing an active, vibrating tillage tool, and it depends non-linearly on the parameters of the soil tillage. Both analytic and numeric procedures describing draught force express this effect. "Draught resistance" was first described by Gorjachkin (1927) in an analytic way. This is a rather simple expression that consists of rational terms

$$F = F_1 + F_2 + F_3 = f \cdot G + k \cdot a \cdot b + \varepsilon \cdot a \cdot b \cdot v^2, \tag{3.1}$$

where $fG = F_1$ is the "no-load" resistance of the plough, which depends on a constant (*f*) and the weight of the plough (*G*). A number of researchers have measured *f*, which may take a value in the range of 0.29–0.5. For example, the modifying factor was established taking into account fast ploughing impacts (Bánházi 1964). Its dependence on the tilling speed is known for different soil types (Sitkei 1968). $kab = F_2$ is the actual "deformation" resistance. Here *a* equals the work depth, *b* is the work width, whereas *k* refers to the deformation resistance factor, which may also vary over a wide range, between 20 and 50 kN m^{-2}! Finally, the third term $\varepsilon abv^2 = F_2$ is the force actually required for moving the furrow-slice, where ε depends on the type of the soil and the shape of the plough, so its value may vary over a wide range: 0.7–12 kNs^2 m^{-4}!

The Gorjachkin relation may be used primarily for the analysis of the draught resistance of normal tillage tools, mainly ploughs. The great advantage of this

L. Fenyvesi (✉) and Z. Hudoba
Hungarian Institute of Agricultural Engineering, Tessedik str. 4, H-2100 Gödöllő, Hungary
e-mail: fenyvesi@fvmmi.hu; hudoba@fvmmi.hu

A.P. Dedousis and T. Bartzanas (eds.), *Soil Engineering*, Soil Biology 20,
DOI 10.1007/978-3-642-03681-1_3, © Springer-Verlag Berlin Heidelberg 2010

method is that it is useful in the analysis of the basic processes, but, on the other hand, it is too general. Due to its universality, the actual value of all the parameters varying over a wide range must be defined.

As far as the numeric methods are concerned, the material equations used (Kushwaha 1998) may be characterised with plastic or elasto-viscoplastic relations. A simple material model cannot be formulated (Chancellor 1994), as the parameters of the model are influenced by too many factors (e.g. humidity, structure), which can only be defined through experiments. A great disadvantage of the calculations carried out with the "infinite element" or "discrete element" method is that they make only the simulation of individual conditions possible.

The periodic nature of the draught force occurring originates from the processes of cutting, compacting and moving the substance (Cooper 1969). This periodicity decreases over a given, critical speed (Stafford 1984). When undisturbed soil is under tillage, the absorbed draught force may be characterised as a random process (Soehne 1956). Simulating the draught force is only possible to a limited extent; no universally valid process characteristics are known. Characterising the measured draught force with statistical methods may make more universal statements possible.

By employing the measured draught force, the power spectral density (PSD) of the draught force may be defined. By defining the diagram, Summers (1984) stated that the delivery of performance is greatest at the frequency of 9.99 Hz; there are smaller maximum values at the frequencies of 995 and 1,187 Hz. It turned out, however, that the features of the performance delivery realised by the draught force are influenced by a number of factors: the type of soil, humidity and the soil condition (Sakai et al. 2005).

The PSD curve is an interesting result (Borsa 1991), which was defined on the basis of the measurements carried out on a wheat stubble field with a medium-deep plough with a very rigid, robust cast-steel beam (Fig. 3.1). Between 0–50 Hz the curve has a monotonic decrease without local extreme values. Therefore the "clean" draught resistance may be described thus: that the delivery of performance occurs between 0–50 Hz in a homogeneous way without a particular frequency. Thus any component of the draught force that may be regarded as periodic may resonate with the natural frequency of the springy tillage tool within the given range.

3.1.2 Active Tillage Tools

We have a number of reasons to assume that decrease of draught force and saving of energy may be achieved by employing an active, vibrating tillage tool:

The internal friction factor of the soils influencing the draught force delivered on the tool decreases if there is a vibration load, as a result of the decrease in the cutting hardness (Savchenko 1958).

The vibration indices of the tillage tools performing a vibrating motion may be coordinated with the indices of the force delivered on the tillage tool, which results in the draught force decreasing in comparison to the fixed tool.

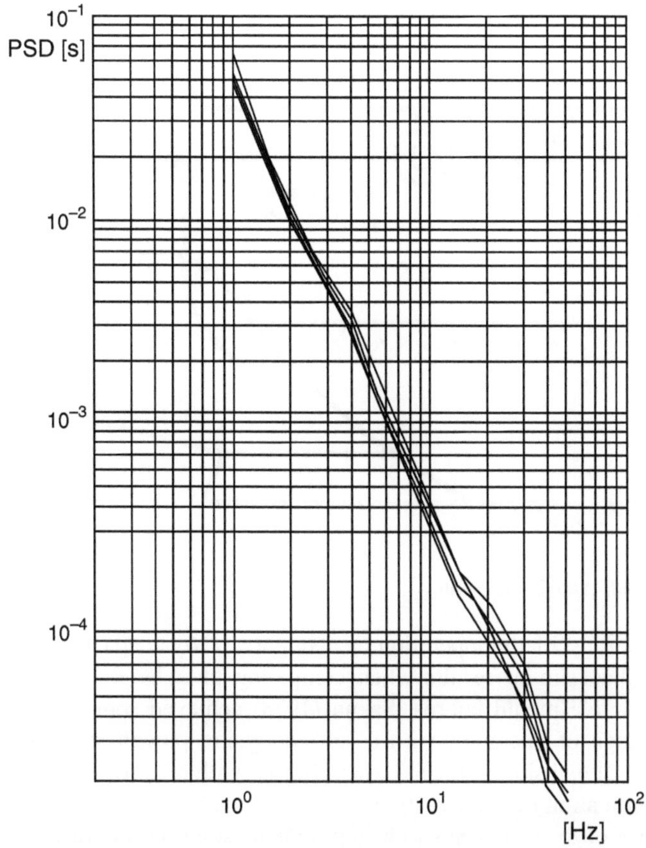

Fig. 3.1 The development of the power spectral density on (compact-soil) land (Borsa 1991)

The maximum force delivered on the tool (e.g. when hitting an obstacle) decreases in proportion to the increase in the time of impulse.

The development of active vibrating tillage tools aims to achieve energy saving by coordinating the effects of the tool and the soil and by decreasing the energy required for the cutting.

The extensive research carried out in this stubble field proved the above advantages in the case of tools based on the employment of a simple edge or wedge, such as looseners (Alexandryan 1963; Totten and Kauffman 1969). This is why vibration is employed in both deep and medium-deep looseners under development (Niyamapa, 1993) and models already on the market.

Generally, the draught force was decreased in the vibrating tillage tools which were developed with the appropriate mechanism and used with an external energy supply; however, this did not result in a decreased energy requirement (Eggenmüller 1958a; Wismer 1968). The total power required for oscillating operation was

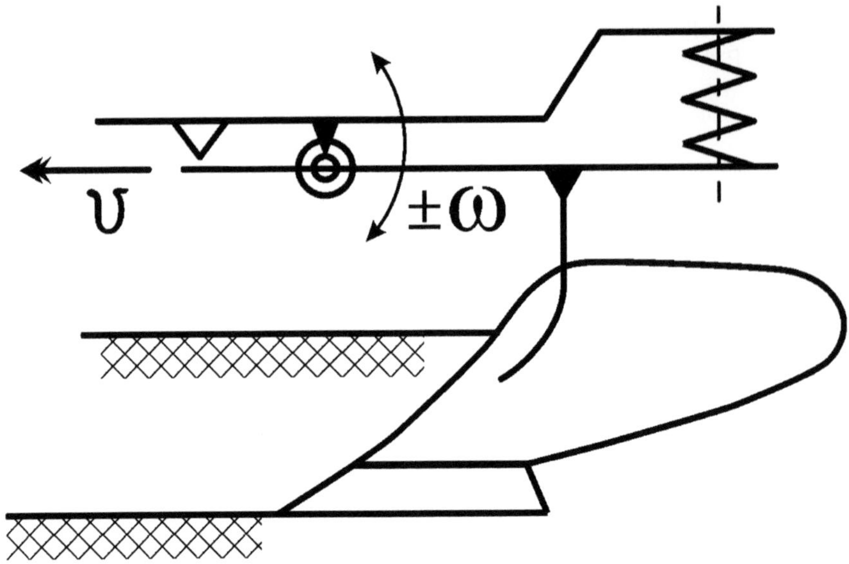

Fig. 3.2 "Sakun" plough with spring

greater for non-oscillating operation by about 40% (Niyamapa and Salokhe 2000). Jóri (1969) had similar results when comparing ploughs equipped with sprung beams and active mould boards. Sakun (1978) achieved some energy saving in a narrow range of operation with a self-energising plough structure (Fig. 3.2). Svercek (1992) had similar results when dealing with a cultivator tool.

The coordination of the vibration of the self-energising sprung tillage tool and the resistance of the soil seems to be a promising solution. In order to develop the equipment for the experiment, the resistance of the soil should be analysed when a fixed (non-vibrating) tillage tool is in use, as this may be regarded as energising the vibrating structure.

3.2 Method and Equipment for the Test

3.2.1 Developing the Test Equipment

According to Eggenmüller (1958b), the individual structural parts of the tillage tool have differing effects when vibrating. Preceding the development of the experimental tillage tools, we constructed a plough on which only the share vibrated, whereas the mouldboard did not. According to the result of the measurements carried out in a soil bin, a perpendicular vibration of the share is more favourable than an edgewise one (Fenyvesi and Mezei 1996). Only the vibration of the share seems to be favourable; however, this is rather difficult to realise on the ploughed field with a device used in normal operational conditions.

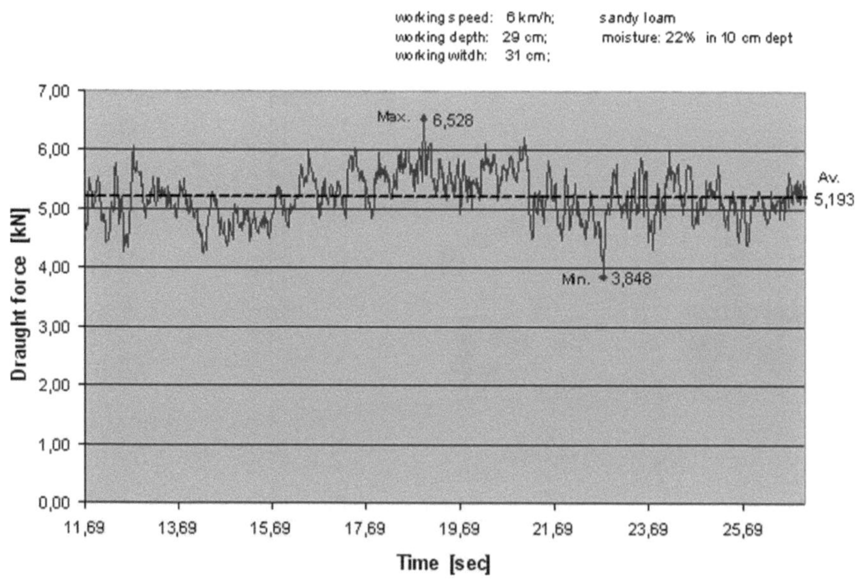

Fig. 3.3 A typical draught force section of the plough

This is why we carried out the modification of tillage tools used in normal operation, and we provided the vibration by means of a spring.

In accordance with Fig. 3.1, a tool had to be developed whose frequency fell in the range of 0–50 Hz, and ideally ranged between 20 and 25 Hz at the basic tooling. As the tillage is carried out on a stubble field (on compact soil), we can assume that there is no special frequency in the domain of variability (Fig. 3.1), so the tool must resonate in the selected domain with the respective draught force components. This may result in saving energy.

However, the given domain of frequency is rather low, which is why a relatively soft spring had to be applied. At the same time, the average value of the draught force (Fig. 3.3) is so high during tillage work that the vibration could not be realised with soft springs.

When developing the test equipment, a mechanism had to be applied that could reduce the effect of the draught force at the spring.

We constructed three versions for the test of the plough and one model for the cultivator. We adapted a plough body with laminated springs that is currently available on the market, by decreasing the number of laminated springs. Thus the plough body performed an oscillating motion against the spring (Fig. 3.4a).

By adapting the plough body, we created a structure that ensured the vibration of the body in the direction of displacement against a cylindrical spring (Fig. 3.4b).

The third solution for the plough ensures the rotating movement of the plough body around an axle, also against cylindrical springs (Fig. 3.5a). The duck-foot shaped cultivator actually has a similar solution to the above (Fig. 3.5b). With these

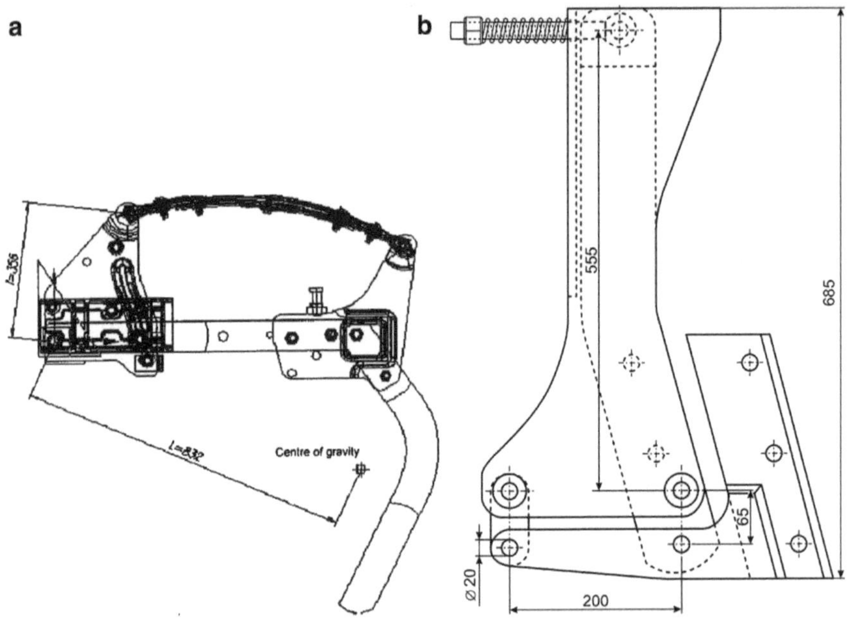

Fig. 3.4 Adapted vibrating ploughs with spring (**a**), and alternating in the direction of motion (**b**)

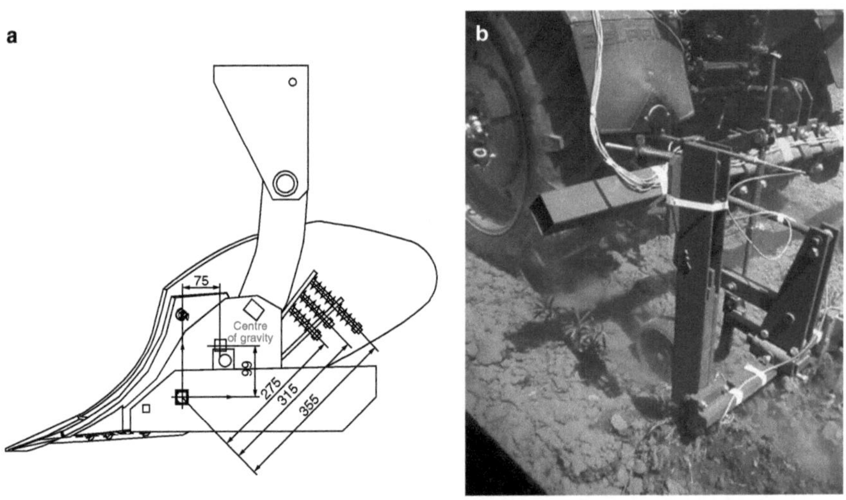

Fig. 3.5 Adapted plough (**a**) and cultivator (**b**) performing alternating rotary movement

solutions for the beam, the nominal geometrical transmission decreases the delivered force to one-tenth, whereas in the case of the cultivator it is decreased to one-fifth of the spring force.

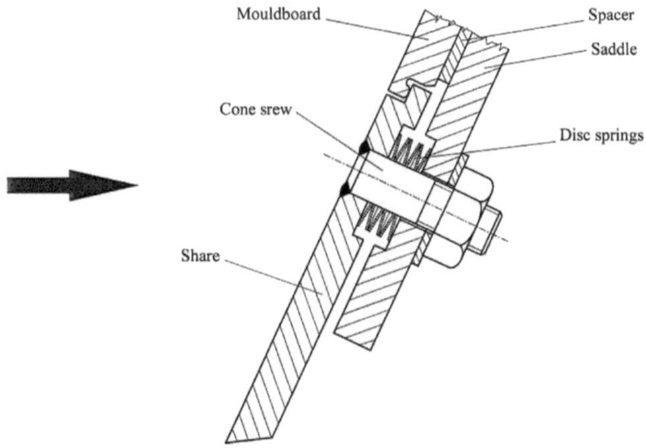

Fig. 3.6 Placing disc springs between the share and the saddle

Fig. 3.7 Mouldboard equipped with disc springs

The equipment introduced above is easy to scale as long as vibration theory is considered; the characteristic frequency is also easily adjustable. However, we have no exact information about the parameters of the forces appearing at certain structural elements of the tool, namely the share and the mouldboard. As long as the cutting function is emphasised in the periodicity of the draught force, it would be practical to ensure the vibration of the share only, with a fixed mouldboard. This solution is also supported by the considerations of rheology.

In this solution, disc springs were installed between the share and the saddle (Fig. 3.6), so that the share could move in a perpendicular direction to the edge. The disc springs are able to absorb large forces with little movement. Changing the number of springs and their method of placement may modify the spring constant (Fig. 3.7).

3.2.2 Defining the Characteristic Frequency of the Test Tools

3.2.2.1 Defining the Characteristic Frequency of the Tillage Tool's Vibration

Natural frequency modelling of the relatively heavy weight system with significant damping taking into consideration the variable effect of the soil is a complicated task.

We employed a simple model for the calculation: with the individual construction we assumed the existence of a clear rotational motion; the moderation and the effect of the soil were neglected. The constructions were imagined as systems of rigid bodies with one degree of freedom. The position of the components of the system were characterised by one parameter, the angle of rotation (φ).

Lagrange's equation of motion may be drawn up for the vibration without moderation (applying the terms of Fig. 3.4a):

$$\frac{\mathrm{d}}{\mathrm{d}t}\frac{\partial E_\mathrm{k}}{\partial \dot{\varphi}} - \frac{\partial E_\mathrm{k}}{\partial \varphi} + \frac{\partial E_\mathrm{s}}{\partial \varphi} = 0, \quad (3.2)$$

where $E_\mathrm{k} = \frac{1}{2}\Theta\dot{\varphi}^2$ (N/m^2) is the kinetic energy; Θ (kg/m^2), the moment of inertia calculated on the pivoting point; $E_\mathrm{s} = \frac{1}{2}\frac{y^2}{c} = \frac{1}{2}\frac{l^2\varphi^2}{c}$ (N/m^2), the static energy accumulated in the springs; y (m), the displacement at the place of the spring; c (m/N), the spring constant; l (m) is the distance of the spring from the pivoting point on a plane in the direction of movement.

$$\text{First term of the Lagrange's equation of motion}: \quad \frac{\mathrm{d}}{\mathrm{d}t}\frac{\partial E_\mathrm{k}}{\partial \dot{\varphi}} = \Theta\ddot{\varphi}, \quad (3.3)$$

$$\text{Second term}: \quad \frac{\partial E_\mathrm{k}}{\partial \varphi} = 0, \quad (3.4)$$

$$\text{Third term}: \quad \frac{\partial E_\mathrm{s}}{\partial \varphi} = \frac{l^2}{c}\varphi. \quad (3.5)$$

Employing the above relations, the equation of motion is:

$$\ddot{\varphi} + \frac{l^2}{\Theta c}\varphi = 0 \quad (3.6)$$

of which the natural angular velocity of the system:

$$\alpha = \sqrt{\frac{l^2}{\Theta c}}\,(1/\mathrm{s}). \quad (3.7)$$

It is clear that the complete structure's mass moment of inertia calculated at the pivoting point, the spring constant and the distance of the spring from the pivoting point are all required in order to calculate the natural frequency of the system. In our example (Fig. 3.4a) the structure's mass moment of inertia calculated at the pivoting point may be replaced by a moment of inertia of a mass point placed at the centre of mass. The AutoCAD drawing software is suitable for defining the centre of mass of plane figures and bodies; this is why the centre of mass of the beam and the plough body in the plane of the direction of motion is available after the construction of the models. L marks the joint mass point of the beam and the body and the distance of the pivoting point in the plane perpendicular to the direction of motion. The structure with the mass of m has its characteristic natural angular velocity:

$$\alpha = \sqrt{\frac{l^2}{mL^2c}}\,(1/\mathrm{s}). \tag{3.8}$$

By employing the relation for the natural frequency, we can draw up the curves of the characteristic cyclical frequency and spring constant for the given solutions (the constructions introduced in Figs. 3.8 and 3.4b are used as examples).

With the help of the relation, the "theoretical" spring characteristics could be defined for the given constructions, taking into consideration the given natural frequency. The stiffness of the spring was amended, starting from the "theoretical" value. The spring stiffness generally had to be increased.

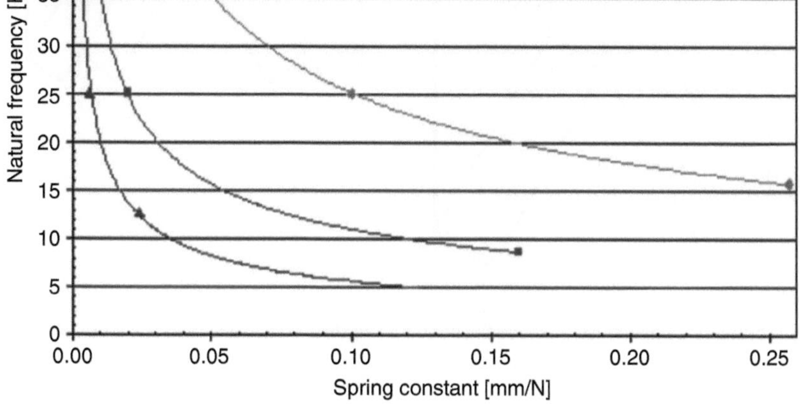

Fig. 3.8 The development of the natural frequency in the function of the spring constant

3.2.2.2 Definition of the Spring Constant of the Structure with Disc Springs

The disc springs were installed with the help of the gripping screws of the share (Fig. 3.6). There were three of them on the test plough, thus the force on each screw is a third of the force delivered on the share. The characteristics of the composition of the springs may be amended by changing the size, number and orientation of the springs (Fig. 3.9).

The spring characteristics calculated and used as a basis for the experimental settings can be seen in case of the various constructions in Fig. 3.10. The disc springs used available on the market met the DIN 2093 standard.

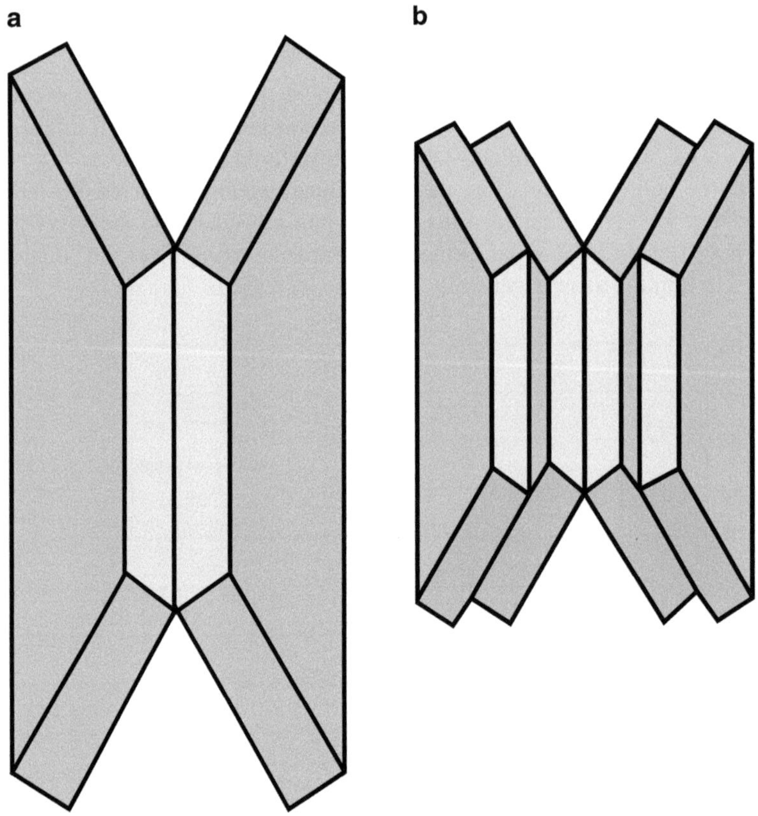

Total force = force of single disc spring Total deflection = 2 x deflection of single disc spring

Total force = 2 x force of single disc spring Total deflection = 2 x deflection of single disc spring

Fig. 3.9 The effect of the method of placing the springs on the parameters of the configuration (www.bellevillesprings.com)

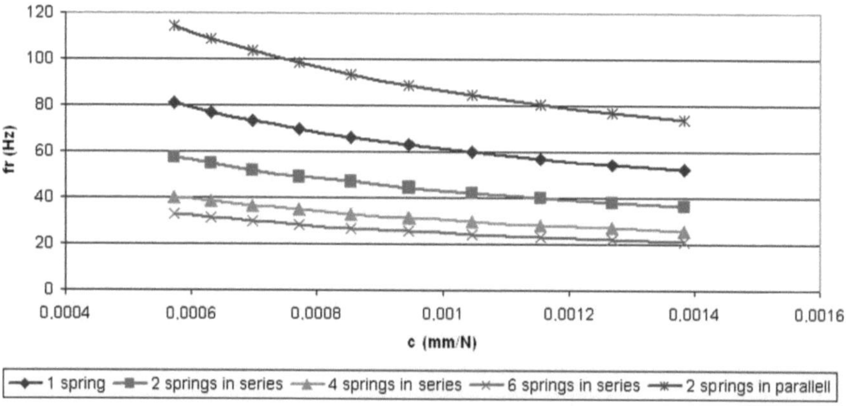

Fig. 3.10 The relation between the calculated spring characteristics for the given disc spring configurations and the calculated natural frequency of the system

3.2.3 Measuring Conditions

Measurements could not be carried out in well-defined conditions, for example in a soil bin as demonstrated above, the draught force depends on the soil characteristics, especially the mechanical conditions. Therefore measurement was carried out on a ploughed field, non-tilled soil and on a vegetated stubble field on the testing site of the Hungarian Institute of Agricultural Engineering (MGI).

The experiments were performed with a three-share, semi-suspended plough; the draught force was measured at the middle and the closing body, using a strain gauge. The measuring points and the connection of the bridge were formed in such a way that it would be only sensitive to the required force component at the appropriate level of sensitivity.

After appropriate preparation, measurement of the vibration in the form of displacement was carried out with an inductive displacement pick-up (Hottinger GmbH). At intervals, the speed was recorded with a Correvit_H (Datron Messtechnik GmbH) speedometer without interruption. The measurement sections were 200–250 m long; we measured in both directions and compared the recorded values. We worked with a MTZ 80 Tractor, and installed the control and vibrating plough bodies at the location of the second and third bodies on the three-bodied LCF-3-35 plough. Due to local effects, we occasionally changed the bodies. No change was carried out with the system equipped with four cultivators.

We installed tensometric measuring points on the bodies and cultivators, sensitive only to the draught force. Prior to that, all measuring points were calibrated either with direct weight-load or with a verified measuring element. We followed suit with the beams of the cultivators. The tillage depth and width were manually sampled at random locations.

We employed the Spider-8 measuring system to amplify and record the signals (Hottinger, Germany), and used statistical methods for processing the signals. We always measured two bodies, installed at the location of either the second or the third bodies. The plough bodies with disc springs could only be measured at the location of the third body.

Connecting the body to the beam with a joint pin ensured that vibration was possible. The counterforce (which also balanced the draught force) was supplied by a spring system on various, variable-force arms whose characteristics could be changed, and which could operate in parallel. Their combination made it possible to create a vibrating motion in the operating range of the plough, the extent of which we could measure with the help of a W50 inductive displacement pick-up. A section of the measured results is illustrated by Fig. 3.11.

In order to ensure the approximate independence of the samples from each other, we used the process of draught force as a scale when setting the sampling interval; this is why we chose one second (Borsa 1988). Thus, in the course of the evaluation, we could employ classical methods of calculation of probability (which assumes having independent samples). A logical exception to this is the spectral analysis, where the sampling distance is set by the conditions of the spectrum estimation. The signals were originally recorded at a frequency of 400 Hz.

The humidity of the medium-compact, sand-clay-soiled stubble field was identical during the compared measurements, as the measuring was practically carried out in one session. The average operational speed was 7 km h^{-1}, the average ploughing depth was 36 cm, and the width was 22 cm. For the duck-foot shaped cultivator, the width was 16 cm, and the depth was 13 cm.

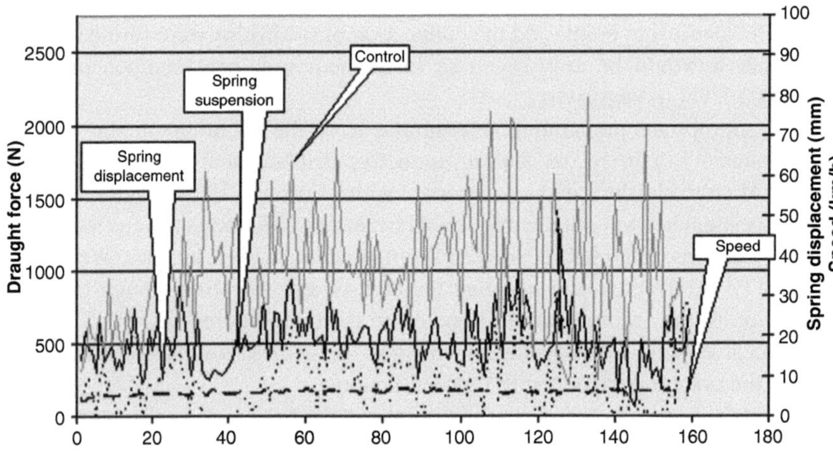

Fig. 3.11 Part of the draught force measured in the vibrating cultivator (medium-compact, sand-clay-soiled stubble field, 13 cm working depth, 7 km h^{-1} speed)

3.2.4 Evaluation Process

The measured signals are randomly variable. In order to evaluate them, mathematical statistical methods must be applied.

The aim of the homogeneity test is to check whether two plough bodies may be regarded as equal regarding draught force. The coincidence of the two distributions is the maximum that can exist between the two random processes in the time range.

Often, rather than supporting the coincidence of the distribution, we would like to point out the source of the difference; namely we would like to see the refutation of the coincidence of the expected values of two processes, as this also means the refutation of the hypothesis of the coincidence of the distribution.

According to our previous tests, the draught force of the plough bodies is not an ergodic process, that is to say, each pull generates a process with different parameters, therefore the comparison is only possible among the results of the same measurement (Borsa 1992). This is what necessitates the comparison of the draught force of two plough bodies installed on the same plough from one and, where possible, a lengthy measurement.

Several methods were used for homogeneity investigation (Graf et al. 1966): Mann-Whitney-Wilcoxon test, Friedman test, Kruska-Wallis's (H) test.

If the result of the homogeneity test shows no contradiction, that is to say, we have no reason to regard the two distributions as different, then it means that the measurements do not support the differences of their parameters either.

Could the main parameter, the expected value of the distribution, be the reason for this difference? In this respect we employed the t or Welch test; in respect of the coincidence of the distribution we employed the F test.

The confidence interval referring to the difference of the two expected values in the case of different distributions is estimated by the approximate method (Sachs 1982), as follows:

$$P[(x_1 - x_2) - t_p^v B \le v_1 - v_2 \le (x_1 - x_2) + t_p^v B] = P. \tag{3.9}$$

where P is the safety; $P = 1 - p$ is the probability of error of the first type (e.g. 5%); t^v is Student's distribution (from Student's distribution table = 1.96 if $v > 300$ and $p = 0.05$); v is degree of freedom and $B = B(s_1,s_2,n)$ value (with identical $n = n_1 = n_2$ sample numbers):

$$v = \frac{(n-1)(s_1^2 + s_2^2)^2}{s_1^4 + s_2^4}, \tag{3.10}$$

$$B = \sqrt{\frac{s_1^2 + s_2^2}{n}}. \tag{3.11}$$

(With the data we gained, the degree of freedom is usually $v > 300$; this is why in the case where the error is $p = 0.05$ it could be taken as $t = 1.96$.)

We employed the normal autocorrelation function to occasionally test the internal relations of the signal (which is required for the selection of the sampling distance), starting from the definition below:

$$R(\tau) = \lim_{T \to \infty} \frac{1}{2T} \int_{-T}^{+T} x(t) \cdot x(t+\tau) \mathrm{d}t \tag{3.12}$$

$$\text{Normal function:} \quad R_{\text{norm}}(\tau) = R^0(\tau) = \frac{R(\tau) - M^2}{D^2}. \tag{3.13}$$

Estimation:

$$R(\tau) \approx \widehat{R}(\tau_k = k \cdot \Delta x) = \frac{1}{n-k} \sum_{i=1}^{n-k} x_i \cdot x_{i+k} \quad k = 0, 1, 2 \ldots, m \tag{3.14}$$

$$R^0(\tau) \approx \widehat{R}^0(\tau_k) = \frac{\widehat{R}(\tau_k) - \bar{x}^2}{s_x^2} \tag{3.15}$$

3.3 Measurement Results

With each plough body we found a setting at which vibration occurred. In these cases we also registered a decrease in the draught force. With respect to the solutions with rotational-vibrational motion, the decrease of draught force is not substantial (Figs. 3.12a and b). When testing the solution with the laminated spring (Fig. 3.4a), we were able to measure a not very significant saving of draught force with four laminates (Fig. 3.12a). With the structure performing rotational motion (Fig. 3.4a), the draught force was most optimal at a spring stiffness of 40–45 N mm^{-1} (Fig. 3.12b). As a result of softening the spring further, the device took a “final state” position and behaved like a fixed device.

The largest decrease in draught force (Fig. 3.13) was achieved where the device was guided in parallel with the draught force (Fig. 3.4b). In this case the optimal value was achieved at a spring stiffness of 4 N mm^{-1}.

With the disc spring version the decrease of draught force was achieved with versions equipped with 7-mm thick, tightly placed 4-spring configurations (Fig. 3.9).

We achieved more spectacular results with the cultivator tool, with a spring stiffness of 3.89 N mm^{-1} (Fig. 3.15) The decrease of draught force is substantial and significant.

Draft force decrease can be shown during the whole measurement range (Fig. 3.15), as well as the high frequency vibration of the tool.

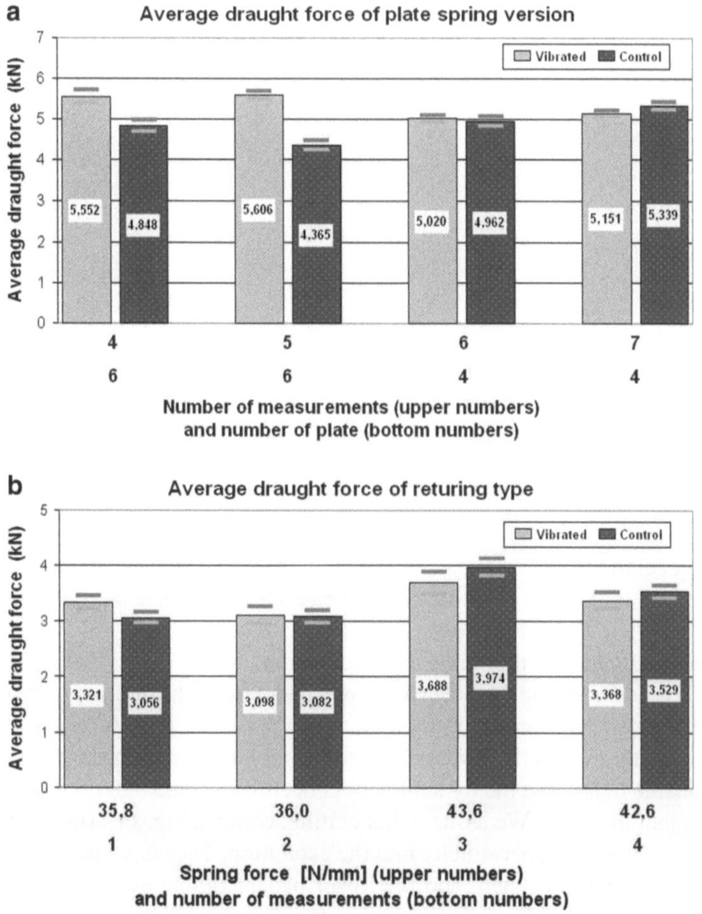

Fig. 3.12 Average draught force with plough structures performing rotational-vibrational motion and with fixed ploughs; (**a**) laminated spring, (**b**) cylindrical spring

The occurrence of draught force in the tilling work is a random process. If we approach this process as the vector sum of many periodical dynamic effects, it is possible that a tillage tool equipped with a spring mechanism will resonate with one of the excitations (draught force components). The real question is how independent is this process from the effects of the other components? Is the decrease of draught force achieved by resonance greater than the loss occurring with the other generating components?

We tried to answer these questions by the measurements carried out with a plough equipped with a laminated spring that is available on the market. As the number of laminates was decreased, so did the draught force (Fig. 3.12a); it fell significantly. The draught force measured with the vibrating plough was practically identical to or slightly less than that of its control (fixed) counterpart.

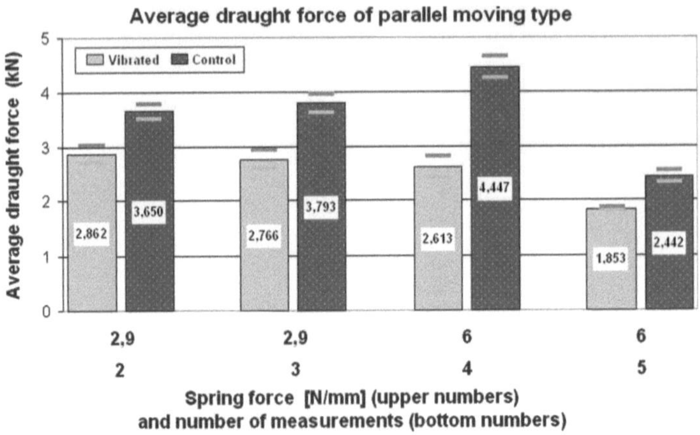

Fig. 3.13 The formation of the average draught force with a plough structure guided in parallel with the draught force, alternating or fixed

Our calculations prove (Sect. 3.2.2) that the "theoretical" characteristic frequency of the tool in the more favourable case (four laminate springs) is less than 50 Hz.

As the frequency of the "draught force components" is also below this value (Fig. 3.1), we can assume the occurrence of a decrease in the draught force resulting from the resonance. In order to prove the assumption, we constructed test devices.

We do not know exactly at which part of the tool the vibration should be applied. We are unable to analyse the dynamic effect occurring at the individual parts of the tool as a random signal. We assume that cutting, compacting, shearing and flow are the processes that bring periodicity into the excitation. Thus it would be appropriate to vibrate primarily the share and the lower part of the mouldboard, whereas the upper part of the mouldboard, performing the carrying of the soil, should not be vibrated. The above theories could only be partly realised in practice (Sect. 3.2.1).

With the plough performing rotational, alternating motion (Fig. 3.5a), insignificant draught force reduction was found at a spring force of 36 N mm^{-1} (Fig. 3.12b). We had the most spectacular results with the plough alternating in the direction of motion (Fig. 3.4b): the draught force fell significantly with spring forces of 2.9 and 6 N mm^{-1}. It looks as if amending the angle of cutting is not favourable; from the point of view of construction the second solution is much more promising. With all the introduced solutions we achieved some decrease in draught force in the expected range, when the natural frequency of the tool was around 10–13 Hz (Fig. 3.8).

Significant decrease in the draught force can also be achieved if only the share is vibrated (Fig. 3.14). We can assume that we did not have a greater decrease because of the amendment of the cutting angle. It is interesting to note that in this case the theoretical natural frequency of the tool is much higher (around 25 Hz) than in the cases above (Fig. 3.10).

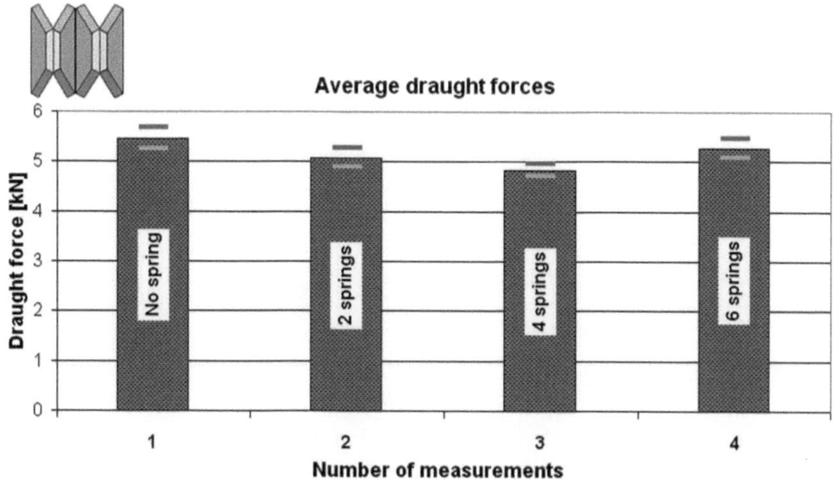

Fig. 3.14 Average values of draught force with 0.7 mm springs at a working speed of 6 km h^{-1}

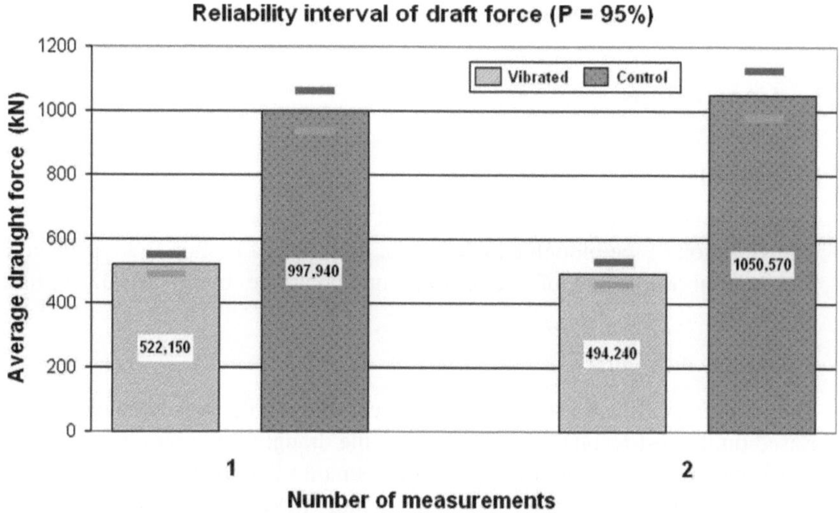

Fig. 3.15 Average values of draught force with vibrating and fixed cultivator tools

As for the vibrating cultivator (Fig. 3.5b), the expected effect of carrying the soil is smaller than with the plough. In the case of a tool whose frequency was set in the region of 15 Hz, we measured a significant decrease in the draught force (Fig. 3.15). This "tool natural frequency" is not identical with the frequency of the tool and resonance of excitation. Having drawn the normal autocorrelation functions of the draught forces measured with fixed and vibrating tools (Fig. 3.16), we can see that resonance is at about 8 Hz. Kerényi et al. (2008) had a similar result modelling the process mathematically and taking into account the moved soil.

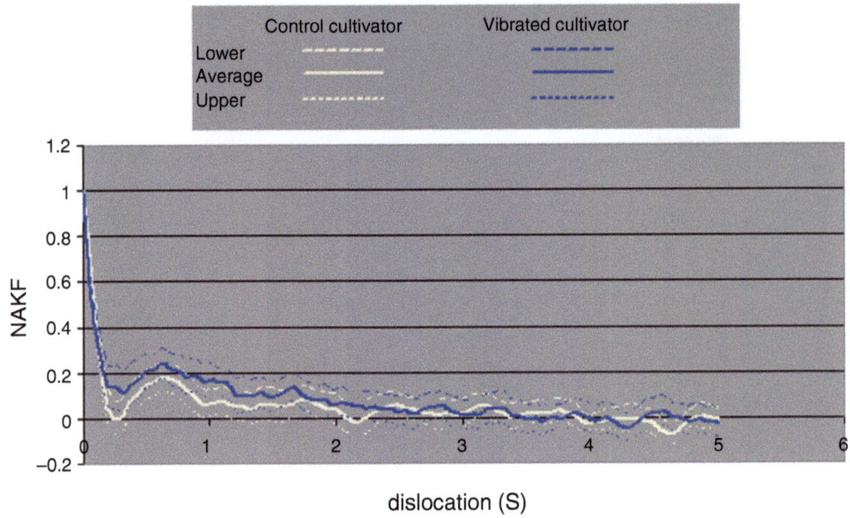

Fig. 3.16 The normal autocorrelation functions of the draught forces that could be measured with fixed (control) and vibrating cultivators (*P* 95%)

3.4 Conclusions

For vibrating ploughs:

- Large changes of the cutting angle are not advantageous.
- The vibration of the plough parallel to the direction of motion is advantageous.
- The natural frequency of the instrument should be up to 50 Hz, optimal 25–30 Hz.

For vibrating cultivator:

- The vibrating cultivator tool works well according to the presented method.
- Based on the test results we can state that the draught force requirement of the experimental vibrating tool is significantly smaller than that of the rigid one.
- To confirm the first results we should make more field tests with a wider range of speeds and different soil types and conditions.

References

Alexandryan KV (1963) The use of vibrations on subsoilers operating in compacted stony soils. Mech Elektrif Sel Hoz Moscow 21(3):49–55

Bánházi J (1964) Talajművelőgépek fejlesztésére irányuló kísérletek (Experiments for development of tillage tools). Publications of the Agricultural University, Gödöllő, Hungary

Borsa B (1988) Bestimmung des Zugkraftsbedarfs von Pflügen aus Messungen als Analyse stochastischer Vorgänge. Grundlagen der Landtechnik, Band 37 2:61–68

Borsa B (1991) Statistical basic properties of plough resistance. Research of Agricultural Engineering. Hungarian Institute of Agricultural Engineering, Gödöllő
Borsa B (1992) Eketestek vontatási ellenállásának stabilitásvizsgálatai: stacionaritás, ergodicitás. (Draught force investigation of plough bottoms: stacionarity, ergodicity). Járművek, Építőipari és Mezőgazdasági Gépek, II: 47–52
Chancellor WJ, Upadhyaya SK, Perumpral JV, Schafer RL, Gill WR, VandenBerg GE (1994) Advances in Soil Dynamics, Vol 1. ASAE Monograph Number 12
Cooper AW (1969) Effects of tillage on soil compaction. Trans. ASAE: 315–364
Eggenmüller A (1958a) Experiments with groups of hoes oscillating in opposite phase. Grandlagen der Landtechnik 10:77–88
Eggenmüller A (1958b) Oscillating implements – kinetics and experiments with single model tools. Grundlagen der Landtechnik 10:55–69
Fenyvesi L, Mezei T (1996) Experiment carried out with active cutting element tillage tool, Hungarian Agricultural Engineering: 1–4
Gorjachkin VP (1927) Teorija pluga, (Theory of plough). Promizdat Publishing House, Moscow
Graf U, Henning HJ, Stange K (1966) Formeln und Tabellen der mathematischen Statistik. Springer, Berlin
Jóri JI (1969) Comparative tests on plough bodies. Research of Agricultural Engineering. Hungarian Institute of Agricultural Engineering, Gödöllő
Kerényi G, Szalai P, Jóri IJ (2008) Vibration analysis and simulation of self vibrating kultivator tool. Gépészet, Budapest, pp 1–6
Kushwaha RL, Shen J (1998) Soil-Machine Interactions, Marcel Dekker, New York, Basel, Hong Kong
Niyamapa T (1993) Soil failure and its application to vibrating tillage tool. Proceedings of the International Conference for Agricultural Machinery Seoul, Korea, pp 1053–1063
Niyamapa T, Salokhe VM (2000) Soil disturbance and force mechanics of vibrating tillage tool. J Terrramech 37:151–166
Sachs L (1982) Statistische Methoden. Fünfte, neubearbeitete Auflage. Springer, Berlin
Sakai KP, Andrade-Sanchez P, Upadhyaya SK (2005) Periodicity and stochastic hierarchical orders of soil cutting force data detected by an "Auto-Regressive Error Distribution Function" (AREF). Trans ADAE V 48(6):2039–2046
Sakun VA (1978) Oputah cnizenia energyemkocty obrabotky pochvu (Possibilities of energy saving in soil tillage), Vect. sz/h Nauky 3. Moscow. pp. 118–130
Savchenko IA (1958) The effect of vibrations on inertial friction in sand. Soil Dynamics, Collection of Papers No. 31., State Publishing House on Construction. Moscow.
Sitkei Gy (1968) Mezőgazdasági gépek talajmechanikai problémái (Soil mechanical problems of agricultural machines). Academic Press, Budapest
Soehne W (1956) Some basic considerations of soil mechanics as applied to agricultural engineering. Grundlagen der Landtechnik 7:11–27
Stafford JV (1984) Force prediction models for brittle and flow failure of soil by draught tillage tools. J Agric Eng Res 29:51–60
Summers ID (1984) Frequency analysis of tillage tool forces. Paper No. 1815 of the Oklahoma Agricultural Experiment Station, p 377–383
Svercek M (1992) Starr oder Abgefedert? Grubberwerzeuge im Verleich. Grundlagen der Landtechnik 3:121–123
Totten DS, Kauffman LC (1969) Dynamic response of a clay soil to an oscillating subsoiler. ASAE Paper No. 69
Wismer RD (1968) Energy applications in tillage and earthmoving. Society of Automative Engineers. Paper No. 680611

Chapter 4
Soil Disturbance and Soil Fragmentation During Tillage

Thomas Keller and Johan Arvidsson

4.1 Introduction

Thirty years ago, it was claimed that one of the principal aims of tillage research was to allow the effects on the soil of using a given tillage implement in given soil conditions to be predicted (Dexter 1979). Although significant advances have been made in soil tillage research, this is still an ambitious aim in soil science (Munkholm et al. 2007).

One of the main reasons for tillage is to break down the soil structure into smaller fragments (i.e. soil fragmentation) for creation of a seedbed. For Sweden and possibly temperate regions elsewhere, Håkansson et al. (2002) reported that a good seedbed for small seeds is obtained when more than 50% of aggregates are smaller than 5 mm. Large aggregates (e.g. aggregates >50 mm) have no agronomic value but can create problems for soil management (Dexter and Bírkas 2004).

Tillage implements exert an external stress on the soil, causing it to fail in some basic modes of soil failure: shear failure, tensile failure and plastic flow (Hadas and Wolf 1983; Hadas 1997; Aluko and Seig 2000). The mode and pattern of soil failure are dependent upon soil conditions (such as soil strength, soil moisture and pre-existing fractures or cracks), and the type and geometry of the tillage implement.

Soil fragmentation only occurs if the soil reaches either shear failure or tensile failure (Hettiaratchi 1988). Fragmentation may be the result of natural processes (e.g. wetting–drying and freezing–thawing cycles) or anthropogenic processes such as tillage. Soil fragmentation due to tillage is also referred to as soil crumbling during tillage.

The following sections discuss soil disturbance and soil fragmentation by tillage. However, the post-tillage evolution of soil structure is not dealt with, nor are long-term effects of tillage systems on soil structure.

T. Keller (✉) and J. Arvidsson
Department of Soil & Environment, SLU, P.O. Box 7014, SE-75007, Uppsala, Sweden
e-mail: thomas.keller@mark.slu.se

A.P. Dedousis and T. Bartzanas (eds.), *Soil Engineering*, Soil Biology 20,
DOI 10.1007/978-3-642-03681-1_4, © Springer-Verlag Berlin Heidelberg 2010

4.2 Soil Disturbance by Tillage

Soil disturbance occurs by shear and/or tensile failure, or by plastic flow (e.g. Aluko and Seig 2000). Tensile failure, also referred to as brittle failure, is the most efficient of these processes, as less energy is required to produce new surface area (Dexter 1988; Díaz-Zorita et al. 2002). In tillage research, shear failure is sometimes also referred to as crescent failure because of the crescent form of shear planes ahead of tines.

The three mechanisms of soil disturbance (i.e. shear failure, tensile failure and plastic flow) are illustrated in Fig. 4.1. In practice, plastic flow (also referred to as lateral failure) mainly occurs for tine implements working below critical depth (Fig. 4.1a and 4.1b), i.e. the depth at which a transition occurs from one type of failure (shear failure) to another (plastic flow). The critical depth is dependent on soil conditions, tine width and rake angle (Spoor and Godwin 1978). Under given soil conditions, the wider the tine, the smaller its rake angle, and the looser the soil surface, the greater the critical depth. As the name suggests, in the case of plastic flow the soil flows around the tine and is not fragmented but is moulded and homogenized, resulting in destruction of soil structure and a reduction in soil strength.

Soil disturbance is dependent upon the shape and geometry of the tillage implement and the initial state of the soil (soil wetness, soil porosity and pre-existing planes of weakness). The complex interactions of implement geometry and initial soil conditions on the mode of soil failure are only partly understood (Hadas 1997). Research on soil–implement interactions to date has mainly focused on soil cutting forces, including draught force requirement (see Chap. 1), rather than soil disturbance.

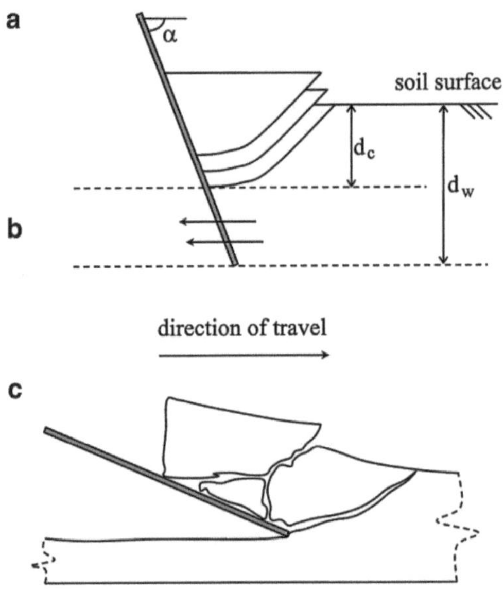

Fig. 4.1 Schematic illustration of basic failure patterns due to tillage implements: (**a**) shear failure; (**b**) plastic flow; (**c**) tensile failure. α: rake angle, d_w: working depth, d_c: critical depth. Adapted from Keller (2004)

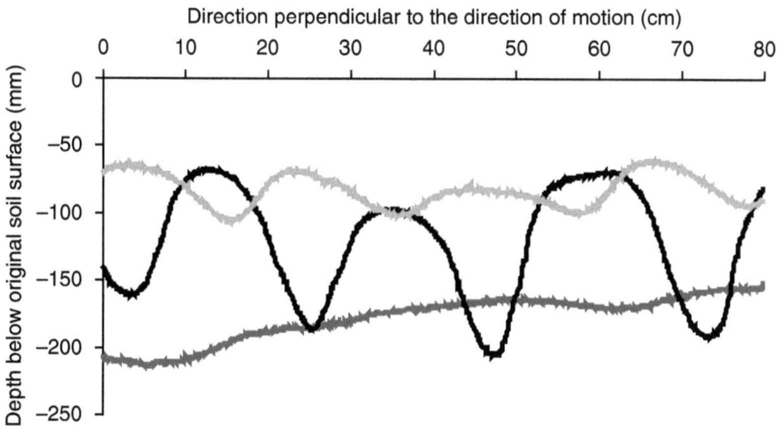

Fig. 4.2 Profile cross-sections of the tilled layer after tillage with a mouldboard plough (*dark grey curve*), a chisel plough (*black curve*) and a disc harrow (*light grey curve*) on a sandy loam. From Keller (2004)

Soil failure patterns and the cross-sectional area of soil disturbance can be quantified by carefully excavating the soil disturbed (loosened) by tillage and measuring the soil relief, e.g. by using a profile meter (Spoor and Godwin 1978), by scanning using laser technology (e.g. Arvidsson and Bölenius 2006) or by using image analysis techniques. Examples of the relief of the bottom of the tilled layer for three different tillage implements are shown in Fig. 4.2. The cross-sectional area of soil disturbance is an important measure when evaluating the efficiency of tillage implements. Specific resistance, i.e. the ratio of draught force requirement to cross-sectional area of soil disturbance, is a better indicator of overall tillage efficiency than draught force itself (Spoor and Godwin 1978; Godwin 2007).

Soil disturbance has been studied for different types of implements and soil conditions in soil bins (e.g. Godwin and Spoor 1977; Onwualu and Watts 1998; Rahman and Chen 2001) and in the field (e.g. Godwin et al. 1984; McKyes and Maswaure 1997; Arvidsson et al. 2004; Kasisira and du Plessis 2006). Extensive geometrical analyses of the contact zones between tillage tool and soil have been reported, particularly for discs, by Godwin et al. (1986), O'Dogherty et al. (1996) and Hettiaratchi (1997a, b). Soil failure patterns have been visualized in studies conducted in glass-sided boxes (e.g. Godwin and Spoor 1977; Fielke 1996; Makanga et al. 1996; Aluko and Seig 2000; Aluko and Chandler 2004). These have led to the conclusion that the mode of soil failure and soil failure pattern are affected by initial soil conditions (soil moisture, soil strength) and implement design parameters such as rake angle. Generally, tensile failure occurs at low rake angles and high soil strength, while shear failure occurs at higher rake angles and lower shear strength (Aluko and Seig 2000).

Calculations and computer simulations of soil disturbance by tillage implements focus on prediction of soil cutting forces rather than failure patterns, and are generally based on the passive earth pressure (also called passive Rankine zone;

Rankine 1857) for analysing retaining walls using the Mohr–Coulomb shear failure criterion (e.g. Payne 1956; Godwin and Spoor 1977; Hettiaratchi 1988; Karmakar and Kushwaha 2006; Tong and Moayad 2006; Godwin and O'Dogherty 2007). Some of the limitations of analytical models based on passive earth pressure theory are reviewed by Karmakar and Kushwaha (2006).

However, soil fragmentation of friable soil relies on tensile or brittle failure (e.g. Aluko and Seig 2000). Consequently, models based on Mohr–Coulomb shear failure are unable to accurately predict soil disturbance and draught force under brittle soil conditions (Aluko and Seig 2000). To our knowledge, no model exists for prediction of mode of soil failure as a function of soil conditions and tillage implement geometry.

4.3 Soil Fragmentation by Tillage

Brittle materials, such as soil in a friable state, fail by propagation of cracks until these join together to form arrays of continuous fracture surfaces (Hallett et al. 1995; Hadas 1997; Aluko and Seig 2000; Dexter and Richard 2009). The soil volume elements delimited by these arrays comprise the soil fragments produced by the tillage. This is consistent with the work of Koolen (1987), Aluko and Koolen (2000), and Snyder and Miller (1989), who attributed soil fragmentation during tillage to failure by cracking of bonds between fragments due to tensile stresses. The pre-existing cracks are also referred to as structural or inter-aggregate pores and are composed mainly of the largest macropores (Hallett et al. 1995).

Failure occurs if the mechanical stress applied overcomes the strength of the material. A property of soil resulting from the concept of soil aggregate hierarchy (Hadas 1987; Dexter 1988) is that as the size of a fragment decreases, its strength increases (Hallett et al. 1995). Hence, the size of the fragments produced by tillage depends upon the applied stress (Díaz-Zorita et al. 2002) and decreases with increasing stress (increasing energy input). Another consequence is that fragmentation results in aggregates with greater strength than the applied stress (Díaz-Zorita et al. 2002).

When fragments are the result of tillage, they are generally called clods (Díaz-Zorita et al. 2002). However, a number of authors use clods to describe large aggregates only. Small aggregates may also be called crumbs. In this chapter, the terms fragment and aggregate are used as interchangeable synonyms.

4.3.1 Soil Friability

Soil friability has been defined as "the tendency of a mass of unconfined soil to disintegrate and crumble under applied stress into a particular size range of smaller aggregates" (Utomo and Dexter 1981; Watts and Dexter 1998).

The concept of friability relies on the theory of brittle fracture and the concept of the "weakest link" (Watts and Dexter 1998). Friability is quantified in terms of the statistical distribution of tensile strength as a function of fragment size (Utomo and Dexter 1981; Watts and Dexter 1998). Measurements of friability can be used to assess the workability of a particular soil, and as an indicator of soil structural conditions and soil physical quality. Chandler and Stafford (1987) proposed a simple field test that estimates soil workability from measurements of the strength and ductility of soil clods.

As concluded by Watts and Dexter (1998), friability increases with increasing aggregate stability, hydraulic conductivity and carbon content, and with decreasing bulk density. Friability is strongly affected by soil water content and reaches a maximum at an intermediate water content close to the plastic limit (Utomo and Dexter 1981; Watts and Dexter 1998).

Future research should investigate the association between values of soil friability prior to tillage and the result of tillage, i.e. the size distribution of aggregates produced by tillage.

4.3.2 Quantifying Soil Fragmentation Produced by Tillage

The result of soil fragmentation by tillage, also referred to as the soil structures produced by tillage, is usually described in terms of the size distribution of aggregates or the fragment mass-size distribution, FSD (Perfect et al. 2002). In the context of tillage, the expression fragment mass-size distribution may be preferred over aggregate size distribution, as aggregate is associated with aggregation, while tillage results in the opposite process, i.e. fragmentation. However, it is commonly accepted that the term aggregate also describes soil structural units resulting from fragmentation (Díaz-Zorita et al. 2002).

It is most common to determine the FSD by sieving and gravimetric analysis, although it is possible to use image analysis techniques (Perfect et al. 2002). Soil is sampled in the field and brought to the laboratory for sieving. In order to facilitate sieving, the freshly tilled soil is usually air-dried prior to determination of FSD (or dry aggregate size distribution, DASD), although sieving of field-moist soil has been reported (e.g. Carter et al. 1998). Note that soil samples must not be allowed to air-dry if the strength of aggregates is to be measured. This is because the strength of soil irreversibly changes when soil is dried to a pore water pressure beyond its driest field condition, i.e. when the effective stresses due to drying overcome the soil pre-shrinkage stress (e.g. Baumgartl and Köck 2004).

From the size distribution of aggregates, the specific surface area of the resulting aggregates, the aggregate mean weight diameter, *MWD*, and the geometric mean diameter, *GMD*, are readily obtained. The specific surface area of aggregates can be expressed as the surface area per unit mass of soil ($m^2\ g^{-1}$), or as the surface area per unit volume of soil ($m^2\ m^{-3}$), $A_{s,V}$:

$$A_{s,V} = A\frac{\rho}{m} \tag{4.1}$$

where ρ is the soil bulk density in g m^{-3}, m is the total mass of the soil in g, and A is the total surface area of all aggregates in m^2 given as (Hadas and Wolf 1983):

$$A = \sum A_i = \sum \frac{6m_i}{\rho\left(\phi_i \phi_{i+1}\right)^{1/2}} \tag{4.2}$$

where A_i is the total surface area of each aggregate size fraction, m_i is the mass of the fraction, and ϕ_i and ϕ_{i+1} are the lower and upper limit, respectively, of aggregate diameter for each size fraction. *MWD* and *GMD* are given as:

$$MWD = \sum p_i \frac{\phi_i + \phi_{i+1}}{2}, \tag{4.3}$$

and

$$GMD = e^{\sum p_i \ln\left(\frac{\phi_i + \phi_{i+1}}{2}\right)} \tag{4.4}$$

where p_i is the proportion by mass of aggregates with diameter between ϕ_i and ϕ_{i+1}. However, the use of any single parameter such as *MWD* to characterize a distribution of fragments is insufficient (Díaz-Zorita et al. 2002).

As demonstrated by Perfect et al. (1993), FSD can be characterized and parameterized using (the cumulative distribution function of) a log-normal distribution, a Rosin–Rammler (1933) distribution or a fractal (power-law) distribution function (Turcotte 1986). Perfect et al. (1993) found that the fractal and Rosin–Rammler functions were theoretically superior to the log-normal function. The Rosin–Rammler distribution is the integral form, i.e. cumulative distribution function, of the two-parameter Weibull (1951) distribution:

$$P(X > x) = 100\mathrm{e}^{-(x/\beta)^{\alpha}} \tag{4.5}$$

where $P\ (X > x)$ is the percentage of aggregates by mass greater than sieve size x, and α and β are constants related to the characteristic size and shape, respectively, of the distribution (Perfect et al. 1993). The Rosin–Rammler distribution is convenient because it has only two parameters and because it is relatively simple. Fractal distribution functions are based on the relationship between fragment diameter, d, and number of fragments, $N(d)$ (Turcotte 1986; Giménez et al. 1998; Perfect et al. 2002):

$$N(d) = k_f d^{-D_f} \tag{4.6a}$$

where k_f is the number of fragments of unit diameter and D_f is the fragmentation fractal dimension. A larger D_f value indicates a FSD dominated by smaller fragments, whereas the value of k_f determines the amount of fragmented material with unit size (Giménez et al. 1998). While it is possible to measure soil fragment numbers, it is more common to determine the fragment mass-size distribution:

$$M(d) = k_{\mathrm{m}} d^{-D_{\mathrm{m}}} \tag{4.6b}$$

where $M(d)$ is the fragment mass, k_{m} is the mass of a fragment of unit diameter and D_{m} is the fractal dimension by mass. Fractal theory and discussion of fractal models for fragmentation of soils are beyond the scope of this chapter. The reader is referred to Turcotte (1986), Perfect and Kay (1995); Perfect (1997), Perfect et al. (2002) and Millán (2004).

4.3.3 Influence of Tillage Implements on Soil Fragmentation

The FSD resulting from tillage has been measured for different tillage implements and different soil conditions by a number of researchers, including Hadas and Wolf (1983), Berntsen and Berre (1993, 2002), Perfect et al. (1993), Larney and Bullock (1994), Giménez et al. (1998), Barzegar et al. (2004), Dexter and Bírkas (2004) and Keller et al. (2007).

Berntsen and Berre (1983) concluded that the FSD after secondary tillage (seedbed preparation) was mainly independent of the implement used and largely determined by the soil state before fracture. This is consistent with the work of Dexter (1979), who concluded that the results of tillage depend much more on soil conditions than on the type of tillage implement used. However, Larney and Bullock (1994), Arvidsson et al. (2004) and Barzegar et al. (2004) found that the FSD produced during tillage was dependent upon both soil wetness and implement type. This is because tillage implements have different effective working depths (cf. Fig. 4.2), rake angles and implement geometry, and hence different specific energy inputs, and because initial soil conditions (water content, bulk density, aggregate density) usually vary over depth. For example, Arvidsson et al. (2004) found increasing soil fragmentation in the order mouldboard plough < chisel plough < disc harrow, which was largely associated with increasing working depth in the order mouldboard plough > chisel plough > disc harrow. This is consistent with findings by Giménez et al. (1998), who reported larger values of k_{m} (4.6b), i.e. denser aggregates, resulting from mouldboard ploughing than from either chisel or disc ploughing, which may relate to inversion of more dense aggregates from deeper in the topsoil layer. Consequently, Giménez et al. (1998) found a lower value of D_{f} (4.6a), i.e. lower fragmentation, after mouldboard ploughing compared with either disc or chisel ploughing.

Although soil fragmentation is obviously affected by tillage tool geometry (e.g. rake angle) and tillage velocity, we are not aware of fundamental studies relating FSD to tillage implement properties.

Studies of (long-term) tillage systems on soil structure have found different results from those cited above. For example, Eghball et al. (1993) reported greater soil fragmentation, i.e. soil dominated by smaller aggregates, indicated by higher D_f values (4.6a) in the order mouldboard plough > chisel > disc > no-till, which is in agreement with Perfect and Blevins (1997). The reason may be differences in secondary tillage between tillage systems (e.g. more intense seedbed preparation in mouldboard tillage systems). Another reason may be that during mouldboard ploughing a greater soil volume is exposed to the weather, with greater effects of wetting–drying and freezing–thawing cycles on post-tillage soil fragmentation.

4.3.4 Influence of Soil Condition on Soil Fragmentation

As stated elsewhere, soil fragmentation is a result of the balance between the (externally) applied stress and the (internal) soil strength. Tillage involves shear failure, tensile failure or plastic flow (and therefore compression), and therefore shear strength, tensile strength or compressive strength may be important. Under field conditions, all three strength components can often play a role simultaneously (e.g. Koolen and Kuipers 1983).

Soil tillage and soil fragmentation are mechanical processes and therefore they are controlled by soil mechanical properties. For example, Koolen and Kuipers (1983) demonstrated that mechanically equal soil conditions yield equal tillage process quantities, such as volume of loosened soil and mean fragment diameter.

Soil mechanical properties including soil strength are dependent upon the soil type and soil conditions, i.e. they are affected by soil texture and organic matter content, the state of compaction (bulk density or void ratio), soil moisture (pore water pressure) and soil structure (e.g. Koolen and Kuipers 1983, Dexter 1988). Of these, soil moisture may be the property that can change most (rapidly) within a given period of time.

Because of the complex nature of different strength components involved in soil fragmentation, because soil moisture controls soil strength, and because soil moisture can change rapidly in a soil, the effects of soil moisture on soil fragmentation are studied most often. Soil fragmentation by tillage is greatly influenced by the soil water content at tillage: if the soil is too wet when tilled, then large clods may be produced and soil structure may be damaged, while if the soil is tilled under too dry conditions, then tillage requires large amounts of energy and large clods may be produced (Dexter and Bird 2001). Therefore, it is important to perform tillage operations at optimal soil moisture conditions, an issue that is discussed further in the following sections.

4.3.5 Modelling Soil Fragmentation Produced by Tillage

Few models exist to predict the structural state of soil following tillage (Perfect et al. 2002).

Dexter (1979) presented empirical equations for the FSD resulting from tillage that predict the statistical distribution of aggregates and voids within the tilled layer. A standard structure is defined for a given soil, water content, working depth, tillage implement and tillage velocity. Factors are then defined which describe deviations from that standard situation, taking into account tillage implement (including tillage depth and number of implement passes), soil management (cropping), water content at tillage and soil compaction. Effects of soil type and tillage velocity on tillage results are not included in the model. Dexter (1979) identified several limitations with his approach, for example that interaction between the different factors could be expected but was not taken into account. Obviously, predicting the FSD resulting from tillage is not an easy task due to complex interactions between soil, soil conditions and implement properties.

Koolen (1977, cited in Perfect et al. 2002) developed a mechanistic model to predict fragments produced by a two-dimensional plough blade in homogeneous cohesive soil. Neither empirical nor mechanistic models have been tested extensively, because empirical models may include site-specific factors, while mechanistic models may require many input parameters because of the complexity of soil–tool interactions (Perfect et al. 2002).

Therefore, models based on probability theory offer an alternative (Hadas 1997; Perfect et al. 2002). Examples of such models include those presented by Perfect (1997) and Perfect et al. (2002). These models estimate the FSD produced by tillage from the pre-tillage soil structure represented as a prefractal porous medium. According to Perfect et al. (2002), the normalized cumulative mass of fragments less than size x, $M^*(x_i < X)$, can be written as:

$$M^*(x_i < X) = \frac{x_i^{-\log P/\log b}}{x_0^{-\log P/\log b} - x_1^{-\log P/\log b}} (x_i \leq x_1) \tag{4.7}$$

where x_i and x_1 are the size of fragments of the ith and first iteration step, respectively, x_0 is the size of the (unfragmented) indicator at $i = 0$, P is the scale-invariant probability of failure, and b is the fractal scale factor, which is related to the normalized size of an element at the ith iteration step, x_i/x_0, by $x_i/x_0 = 1/b^i$. Perfect et al. (2002) showed that (4.7) was useful for parameterizing the FSD produced by a specific energy input. However, they did not estimate the constants in (4.7) (i.e. x_0, b and P) independently but fitted the equation to measured data. Further research is needed in order to obtain independent estimates of x_0, b and P in order to predict the FSD resulting from a particular tillage operation on a particular soil under given soil conditions.

Recent work by Dexter and co-workers (Dexter and Bird 2001; Dexter 2004b; Dexter and Bírkas 2004; Dexter et al. 2005; Keller et al. 2007; Dexter and Richard 2009) has explored the relationship between soil fragmentation by tillage and the water retention characteristics of soil. Keller et al. (2007) showed that the aggregate size distribution produced by tillage at the optimum water content for tillage can be predicted from the value of S (Dexter 2004a, b, c). This is discussed further in the following sections.

4.3.6 Optimum Water Content for Tillage

The optimum water content for tillage, w_{opt}, has been defined as the water content at which production of small aggregates is greatest, production of large aggregates is least (Dexter and Bird 2001), and the specific surface area of the aggregates produced is largest (Keller et al. 2007).

The physical explanation for the existence of an optimum soil water status for tillage is based on the assumption that soil fragmentation due to tillage occurs by the propagation of pre-existing cracks (Hallett et al. 1995; Hadas 1997; Dexter and Bird 2001; Dexter and Richard 2009). The ability of such cracks to elongate or expand under mechanical stress is greatest when they are air-filled. This is achieved by either reduction of air pressure within the cracks, or by inflow of air from surrounding air-filled pores. In contrast, when soil is wetter than the optimum, these micro-cracks contain water that cannot readily expand and cannot rapidly flow from surrounding pores to allow crack growth (Dexter and Richard 2009). In other words, tillage produces coarser soil fragments when soil cracks are less able to elongate under mechanical stress. For soil wetter than the optimum, this occurs because of the inability of water-filled cracks to expand, while for soil drier than the optimum, it occurs because of the increased strength of the soil between the cracks. Hence, the optimum water status for tillage is the result of a balance between these two opposing effects (Dexter and Richard 2009).

Koolen (1987) and Aluko and Koolen (2000) proposed a capillary crumbling model, which is based on a capillary bonding stress, p_w:

$$p_w = u\chi \tag{4.8}$$

where u is the soil water pressure and χ the degree of pore saturation. Koolen (1978) reasoned that p_w as a function of u and χ within aggregates (intra-aggregate) would differ from that between aggregates (inter-aggregate), as shown in Fig. 4.3. According to Mullins and Panayiotopolous (1984); Snyder and Miller (1989) and Aluko and Koolen (2000), only the fraction of water in the inter-aggregate pore space contributes to the strength of a structured soil. Consequently, (4.8) can be written in terms of the tensile strength, σ_t (Aluko and Koolen 2000):

$$\sigma_t = u\chi_i \tag{4.9}$$

where χ_i is the degree of inter-aggregate pore saturation. While p_w is an expression of the bonding strength of the bulk soil, σ_t is an expression of the bonding strength at the inter-aggregate locations (Aluko and Koolen 2000). Hence, with regard to (4.9) and Fig. 4.3, w_{opt} is the water content at which the force needed for crumbling is minimal, i.e. the water content at which $\sigma_t = u\chi_i$ reaches a minimum.

The optimum water content for tillage has been reported to be slightly below the lower plastic (or lower Atterberg) limit, *PL* (e.g. Ojeniyi and Dexter 1979; Müller et al. 2003; Barzegar et al. 2004). This is in agreement with Watts and Dexter

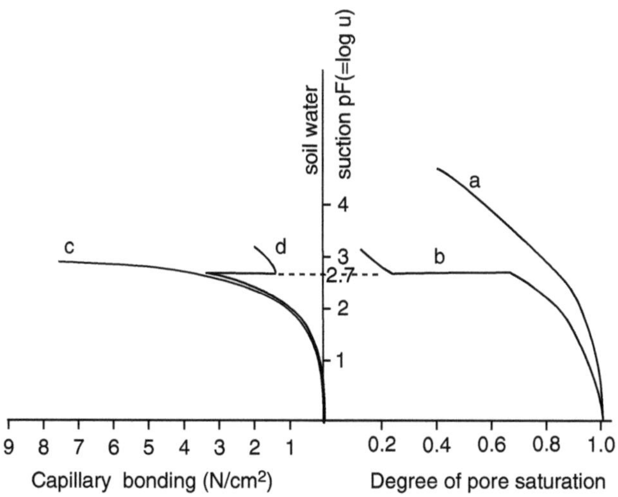

Fig. 4.3 The p_w-u-χ diagram of capillary crumbling proposed by Koolen (1987). Curves: (**a**) pF (log*u*) curve within aggregates (intra-aggregate); (**b**) pF curve between aggregates (inter-aggregate); (**c**) capillary bonding stress within aggregates (intra-aggregate); (**d**) capillary bonding stress between aggregates (inter-aggregate). From Aluko and Koolen (2000) with permission from Elsevier

(1998), who reported that friability (see Sect. 4.3.1) reaches a maximum at water contents around *PL*. However, while *PL* is determined on moulded soil, tillage is carried out on undisturbed soil, which can have a high structural porosity, and therefore there are problems with the use of *PL* (Dexter and Bird 2001; Dexter and Richard 2009).

Dexter and Bird (2001) defined w_{opt} from the water retention characteristics of soil and found w_{opt} to be equal to the water content at the inflection point of the water retention curve, w_{infl}, when the water retention curve is plotted as the natural logarithm of the water suction, *h*, against gravimetric water content, *w*. The work of Dexter and Bird (2001) is supported by experimental data on Hungarian soils (Dexter and Birkás 2004) and Swedish soils (Keller et al. 2007). Keller et al. (2007) showed that w_{infl} provides a better estimate of w_{opt} than any proportion of *PL*.

Dexter and Bird (2001) and Dexter et al. (2005) also defined a lower (dry) and an upper (wet) tillage limit that can be calculated from parameters of the water retention curve. Tillage can be satisfactorily performed at water contents within the lower and upper tillage limit. However, the tillage limits defined are somewhat arbitrary and to our knowledge they have not been tested.

Dexter and Bird (2001), Dexter and Bírkas (2004), Dexter et al. (2005) and Keller et al. (2007) used the van Genuchten (1980) equation for parameterizing the soil water retention curve. The van Genuchten equation is unimodal (or monomodal), i.e. it has exactly one inflection point. However, most soils have pore size distributions that are bimodal (Kutílek 2004; Dexter et al. 2008). This means that the water retention curve has two inflection points, representing two distinct peaks of pore size corresponding to the textural and structural pore spaces. Dexter and Richard (2009) re-examined experimental data published by Dexter and Birkás (2004) and Keller et al. (2007) using the bimodal, double-exponential water retention equation of Dexter et al. (2008) and showed that h at the inflection point of the van Genuchten (1980) equation was between the values of h_1 and h_2 corresponding to the peak of the matrix and structural pore space, respectively. That is, at w_{opt} the structural pore space is mostly drained (air-filled), whereas the matrix pore space is mostly water-filled. Dexter and Richard (2009) further concluded that, for the time being, the double-exponential equation does not provide better estimates of w_{opt} than the van Genuchten (1980) equation.

4.3.6.1 Size Distribution of Aggregates Produced by Tillage at w_{opt}

Dexter and Birkás (2004) and Keller et al. (2007) showed that the amount of aggregates of different sizes, and hence the FSD resulting from tillage at w_{opt}, can be predicted from the S value defined by Dexter (2004a, b, c). The value of S is equal to the slope of the water retention curve at its inflection point, when the water retention curve is plotted as log$h$ against w. It can be used as an index of soil physical quality as shown by Dexter (2004a, b, c). Keller et al. (2007) found that the specific surface area (4.1 and 4.2) of the aggregates produced by tillage is proportional to the value of S. The aggregate size distribution for a specific soil can be estimated from the value of S as shown in Fig. 4.4. Unfortunately, Dexter and Birkás (2004) and Keller et al. (2007) only made predictions for w_{opt} and not for other degrees of soil wetness. Furthermore, tillage was performed with a mouldboard

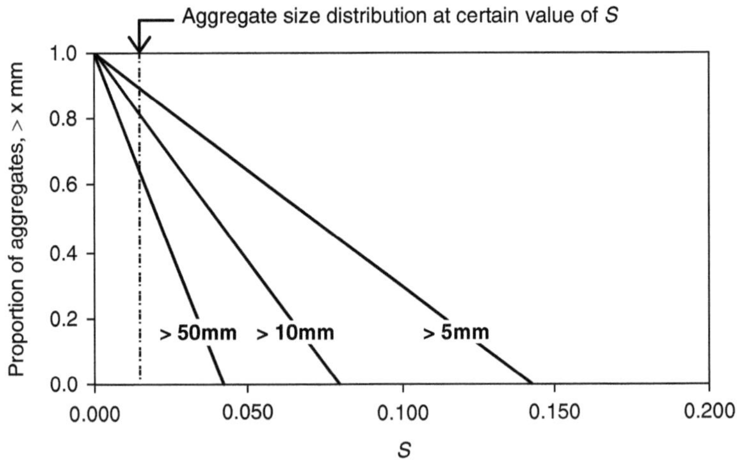

Fig. 4.4 Proportions by mass of aggregates produced by mouldboard ploughing at the optimum water content for tillage. From Keller et al. (2007) with permission from Elsevier

plough in both studies and hence generalizations cannot be made for other tillage implements.

Dexter and Richard (2009) found that the value of S is a function of both the structural and textural pore space. This explains why both lower bulk density and higher organic matter content are associated with good soil physical quality, even though, as reported by Dexter et al. (2008), compaction affects the structural pore space and organic matter affects the textural pore space.

While the approach of predicting the FSD from the soil water retention characteristics adopted by Dexter and Birkás (2004) and Keller et al. (2007) is attractive, the physical mechanisms involved are not fully understood (Dexter and Richard 2009).

4.4 Conclusions

This chapter provides an overview of soil tillage research on soil disturbance and soil fragmentation. Numerous studies on the subject have been reported, but only a few prediction models exist. This is due to the complex interactions of implement geometry and soil conditions on soil failure and soil fragmentation. There is an apparent inconsistency in that models of soil disturbance and soil cutting forces rely on Mohr–Coulomb shear failure, while the theory of soil fragmentation is based on the mechanics of crack propagation in brittle materials. Furthermore, studies on soil disturbance and soil cutting forces seldom include soil fragmentation and vice versa.

Research into the fundamental reasons behind soil failure, as affected by soil consistency, is required. For example, further studies are needed to refine models for prediction of the fragment mass-size distribution produced by tillage as a function of the type of tillage implement and the soil initial state. Future research should also investigate the association between soil break-up and soil fragmentation.

References

Aluko OB, Chandler HW (2004) Characterisation and modelling of brittle fracture in two-dimensional soil cutting. Biosys Engng 88:369–381

Aluko OB, Koolen AJ (2000) The essential mechanics of capillary crumbling of structured agricultural soils. Soil Tillage Res 55:117–126

Aluko OB, Seig DA (2000) An experimental investigation of the characteristics of and conditions for brittle failure in two-dimensional soil cutting. Soil Tillage Res 57:143–157

Arvidsson J, Bölenius E (2006) Effects of soil water content during primary tillage – laser measurements of soil surface changes. Soil Tillage Res 90:222–229

Arvidsson J, Keller T, Gustafsson K (2004) Specific draught for mouldboard plough, chisel plough and disc harrow at different water contents. Soil Tillage Res 79:221–231

Barzegar AR, Hashemi AM, Herbert SJ, Asoodar MA (2004) Interactive effects of tillage system and soil water content on aggregate size distribution for seedbed preparation in Fluvisols in southwest Iran. Soil Tillage Res 78:45–52
Baumgartl T, Köck B (2004) Modeling volume change and mechanical properties with hydraulic models. Soil Sci Soc Am J 68:57–65
Berntsen R, Berre B (1993) Fracturing of soil clods and the soil crumbling effectiveness of draught tillage implements. Soil Tillage Res 28:79–94
Berntsen R, Berre B (2002) Soil fragmentation and the efficiency of tillage implements. Soil Tillage Res 64:137–147
Carter MR, Sanderson JB, MacLeod JA (1998) Influence of time of tillage on soil physical attributes in potato rotations in Prince Edward Island. Soil Tillage Res 49:127–137
Chandler HW, Stafford JV (1987) A simple field test for predicting ease of cultivation. Soil Tillage Res 10:269–275
Dexter AR (1979) Prediction of soil structures produced by tillage. J Terramechanics 16:117–127
Dexter AR (1988) Advances in characterization of soil structure. Soil Tillage Res 11:199–238
Dexter AR (2004a) Soil physical quality. Part I: theory, effects of soil texture, density and organic matter, and effects on root growth. Geoderma 120:201–214
Dexter AR (2004b) Soil physical quality: Part II. friability, tillage, tilth and hard-setting. Geoderma 120:215–225
Dexter AR (2004c) Soil physical quality: Part III: unsaturated hydraulic conductivity and general conclusions about S-theory. Geoderma 227-239
Dexter AR, Bird NRA (2001) Methods for predicting the optimum and the range of water contents for tillage based on the water retention curve. Soil Tillage Res 57:203–212
Dexter AR, Birkás M (2004) Prediction of the soil structures produced by tillage. Soil Tillage Res 79:233–238
Dexter AR, Richard G (2009) Tillage of soils in relation to their bi-modal pore size distribution. Soil Tillage Res 103:113–118
Dexter AR, Czyz EA, Bikás M, Diaz-Pereira E, Dumitru E, Enache R, Fleige H, Horn R, Rajkaj K, de la Rosa D, Simota C (2005) SIDASS project, Part 3: the optimum and range of water contents for tillage – further developments. Soil Tillage Res 82:29–37
Dexter AR, Czyz EA, Richard G, Reszkowska A (2008) A user-friendly water retention function that takes account of the textural and structural pore spaces in soil. Geoderma 143:243–253
Díaz-Zorita M, Perfect E, Grove JH (2002) Disruptive methods for assessing soil structure. Soil Tillage Res 64:3–22
Eghball B, Mielke LN, Calvo GA, Wilhelm WW (1993) Fractal description of soil fragmentation for various tillage methods and crop sequences. Soil Sci Soc Am J 57:1337–1341
Fielke JM (1996) Interactions of the cutting edge of tillage implements with soil. J Agric Engng Res 63:61–72
Giménez D, Allmaras RR, Huggins DR, Nater EA (1998) Mass, surface, and fragmentation fractal dimensions of soil fragments produced by tillage. Geoderma 86:261–278
Godwin RJ (2007) A review of the effect of implement geometry on soil failure and implement forces. Soil Tillage Res 97:331–340
Godwin RJ, O'Dogherty MJ (2007) Integrated soil tillage force prediction models. J Terramechanics 44:3–14
Godwin RJ, Spoor G (1977) Soil failure with narrow tines. J Agric Engng Res 22:213–228
Godwin RJ, Seig DA, Allott M (1986) Soil failure and force prediction for soil engaging discs. Soil Use Manage 3:106–114
Godwin RJ, Spoor G, Soomro MS (1984) The effect of tine arrangement on soil forces and disturbance. J Agric Engng Res 30:47–56
Hadas A (1987) Long-term tillage practice effects on soil aggregation modes and strength. Soil Sci Soc Am J 51:191–197
Hadas A (1997) Soil tilth – the desired soil structural state obtained through proper soil fragmentation and reorientation processes. Soil Tillage Res 43:7–40

Hadas A, Wolf D (1983) Energy efficiency in tilling dry clod-forming soils. Soil Tillage Res 3:47–59
Håkansson I, Myrbeck Å, Etana A (2002) A review of research on seedbed preparation for small grains in Sweden. Soil Tillage Res 64:23–40
Hallett PD, Dexter AR, Seville JPK (1995) Identification of pre-existing cracks on soil fracture surfaces using dye. Soil Tillage Res 33:163–184
Hettiaratchi DRP (1988) Theoretical soil mechanics and implement design. Soil Tillage Res 11:325–347
Hettiaratchi DRP (1997a) The soil contact zones of concave agricultural discs: Part 1, Theoretical analysis. J Agric Engng Res 67:113–125
Hettiaratchi DRP (1997b) The soil contact zones of concave agricultural discs: Part 2, Calculation procedures. J Agric Engng Res 67:127–139
Karmakar S, Kushwaha RL (2006) Dynamic modeling of soil-tool interaction: An overview from a fluid flow perspective. J Terramechanics 43:411–425
Kasisira LL, du Plessis HLM (2006) Energy optimization for subsoilers in tandem in a sandy clay loam soil. Soil Tillage Res 86:185–198
Keller T (2004) Soil Compaction and Soil Tillage – Studies in Agricultural Soil Mechanics. PhD Thesis. Agraria 489, Swedish University of Agricultural Sciences, Uppsala.
Keller T, Arvidsson J, Dexter AR (2007) Soil structures produced by tillage as affected by soil water content and the physical quality of soil. Soil Tillage Res 92:45–52
Koolen AJ (1977) Soil Loosening Processes in Tillage: Analysis, Systematics and Predictability. PhD-Thesis. Mededelingen Landbouwhogeschool 77-17, Wageningen The Netherlands.
Koolen AJ (1987) Workability Limits and their Soil Physical Background. PAGV-Verslag No 64. Lelystad, The Netherlands, pp 21-41.
Koolen AJ, Kuipers H (1983) Agricultural Soil Mechanics. Advanced Series in Agricultural Sciences 13, Springer-Verlag, Berlin Heidelberg New York Tokyo.
Kutílek M (2004) Soil hydraulic properties as related to soil structure. Soil Tillage Res 79:175–184
Larney FJ, Bullock MS (1994) Influence of soil wetness at time of tillage and tillage implement on soil properties affecting wind erosion. Soil Tillage Res 29:83–95
Makanga JT, Salokhe VM, Gee-Clough D (1996) Effect of rake angle and aspect ratio on soil failure patterns in dry loam soil. J Terramechanics 33:233–252
McKyes E, Maswaure J (1997) Effect of design parameters of flat tillage tools on loosening of a clay soil. Soil Tillage Res 43:195–204
Millán H (2004) Fragmentation of soil initiators: application of the pore-solid fractal model. Fractals 12:357–363
Müller L, Schindler U, Fausey NR, Lal R (2003) Comparison of methods for estimating maximum soil water content for optimum workability. Soil Tillage Res 72:9–20
Mullins CE, Panayiotopolous KP (1984) The strength of unsaturated mixtures of sand and kaolin and the concept of effective stress. J Soil Sci 35:459–468
Munkholm LJ, Perfect E, Grove J (2007) Incorporation of water content in the Weibull model for soil aggregate strength. Soil Sci Soc Am J 71:682–691
O'Dogherty MJ, Godwin RJ, Hann MJ, Al-Ghazal AA (1996) A geometrical analysis of inclined and tilted spherical plough discs. J Agric Engng Res 63:205–218
Ojeniyi SO, Dexter AR (1979) Soil factors affecting the macrostructures produced by tillage. Trans ASAE 22:339–343
Onwualu AP, Watts KC (1998) Draught and vertical forces obtained from dynamic soil cutting by plane tillage tools. Soil Tillage Res 48:239–253
Payne PCJ (1956) The relationship between the mechanical properties of soil and the performance of simple cultivation implements. J Agric Engng Res 1:23–50
Perfect E (1997) Fractal models for the fragmentation of rocks and soils: A review. Eng Geol 48:185–198
Perfect E, Blevins RL (1997) Fractal characterization of soil aggregation and fragmentation as influenced by tillage treatment. Soil Sci Soc Am J 61:896–900

Perfect E, Kay BD (1995) Applications of fractals in soil and tillage research: A review. Soil Tillage Res 36:1–20
Perfect E, Díaz-Zorita M, Grove JH (2002) A prefractal model for predicting soil fragment mass-size distributions. Soil Tillage Res 64:79–90
Perfect E, Kay BD, Ferguson JA, da Silva AP, Denholm KA (1993) Comparison of functions for characterizing the dry aggregate size distribution of tilled soil. Soil Tillage Res 28:123–139
Rahman S, Chen Y (2001) Laboratory investigation of cutting forces and soil disturbance resulting from different manure incorporation tools in a loamy sand soil. Soil Tillage Res 58:19–29
Rankine W (1857) On the stability of loose earth. Philosophical Transactions of the Royal Society of London, vol 147.
Rosin P, Rammler E (1933) The laws governing the fineness of broken coal. J Inst Fuel 7:29–36
Snyder VA, Miller RD (1989) Soil deformation and fracture under tensile forces. In: Larson WE, Gupta SC (eds) Mechanics and Related Processes in Structured Agricultural Soils. NATO ASI Series E, Kluwer, vol 172. Academic Publishers, Dordrecht, pp 22-35.
Spoor G, Godwin RJ (1978) An experimental investigation into the loosening of soil by rigid tines. J Agric Engng Res 23:243–257
Tong J, Moayad BZ (2006) Effects of rake angle of chisel plough on soil cutting factors and power requirement: a computer simulation. Soil Tillage Res 88:55–64
Turcotte DL (1986) Fractals and fragmentation. J Geophys Res 91:1921–1926
Utomo WH, Dexter AR (1981) Soil friability. J Soil Sci 32:203–213
van Genuchten MT (1980) A closed-form equation for predicting the hydraulic conductivity of unsaturated soils. Soil Sci Soc Am J 44:892–898
Watts CW, Dexter AR (1998) Soil friability: Theory, measurement and the effects of management and organic carbon content. Eur J Soil Sci 49:73–84
Weibull W (1951) A statistical distribution function of wide applicability. J Appl Mech 18:293–297

Part II
Soil Dynamics Traffic and Traction

Chapter 5
Robotics and Sustainability in Soil Engineering

S. Fountas, T.A. Gemtos, and S. Blackmore

5.1 Introduction

Agricultural mechanization started with agriculture about 10,000 years ago. Humans used initially tools (such as stone sickles for cereal harvesting) to assist in their work. Animal power helped man to reduce the labour requirements for crop and animal production. Mechanical power use has greatly increased from the beginning of the twentieth century with the use of internal combustion engines in tractors. From that period until today, larger tractors have been employed to enhance mechanization and reduce labour requirements and costs. The mean tractor power and weight increased throughout this period, although the weight per unit of power has reduced (Goering 1992). The increased weight of the tractors increased the load on the soil causing increased compaction problems. Larger tractors increased the stresses imposed on the soil to develop traction during field work, increased soil deformation and increased soil structure destruction.

Developed agriculture uses many types of machinery to enhance production. Most of these machines are adaptations from older designs and some have not changed much for centuries. During the industrial revolution, new energy sources became available that allowed machines to replace human labour. The interaction between machines and operators has hardly changed since that time. The human operators use their intelligence to operate a mechanical tractor and implement. Automatic sub-systems are now becoming commonplace and new driver-assist technologies are now being commercialized, such as implement position during

S. Fountas (✉) and T.A. Gemtos
Laboratory of Farm Mechanisation, Department of AgricultureUniversity of Thessaly, Crop Production and Rural Environment, Fytoko Street, N. Ionia 38446, Magnesia, Greece
e-mail: sfountas@uth.gr

S. Blackmore
Center for Research and Technology, Thessaly (CERETETH), 1st Industrial zone, Volos 38500, Greece

A.P. Dedousis and T. Bartzanas (eds.), *Soil Engineering*, Soil Biology 20,
DOI 10.1007/978-3-642-03681-1_5, © Springer-Verlag Berlin Heidelberg 2010

field work through the hydraulic hitch system, and straight line assisted steering. It would seem inevitable that machines will become smarter in the future, to the point where they are able to carry out the tasks autonomously.

Autonomous vehicles have been widely used in industrial production and warehouses, where a controlled environment can be guaranteed. In agriculture, research into driverless vehicles has always been a dream but serious research started in the early 1960s. These projects mainly involved automatically steered tractors (a review can be seen at Wilson 2000). Furthermore, Hollingum (1999) reviewed the agricultural robotic developments around the world and Kondo and Ting (1998) elaborated on robotics for bio-production systems, including open fields. In recent years the development of autonomous vehicles in agriculture has experienced an increased interest. There are a number of prototypes that have been reported in horticultural crops, such as oranges (Hannan and Burks 2004), apples (Kataoka et al. 2001), strawberries (Kondo et al. 2005), and tomatoes (Chi and Ling, 2004). For field crops there are also a number of prototypes, such as the Demeter system for automated harvesting equipped with a video camera and GPS for navigation (Pilarski et al. 2002), the autonomous Christmas tree weeder (Have et al. 2002), and the API platform for patch spraying (Bak and Jakobsen 2003).

5.1.1 Current Mechanization Trends

Most new machines brought to the market are bigger than the previous model. From discussing this issue with equipment manufacturers, this trend is likely to continue into the future. The driving force for this growth is to take advantage of the economies of scale that larger machines bring with them. This can be easily seen if the cost of the operator is taken into account. As most operators are paid by the hour, a larger machine that can increase the work rate over a smaller one can have a significant economic advantage.

This size increase does not only bring benefits. Large machines are only viable when working in large fields, as turning, positioning and transport are all non-productive activities. Although many farms have removed field boundaries to take advantage of the larger machines, many smaller farms cannot follow suit due to environmental concerns, and suffer economically because of it.

As the equipment becomes larger, it also becomes very capital-intensive with new tractors and combines becoming prohibitively expensive for the small and medium-sized farm. Reliability also becomes an issue, as most farms have only one large tractor and all processes are carried out in series. If one part of the mechanization system breaks down, then all field operations stop.Even with the very high work rates of large machines, any failure during the critical period can cause failures to establish crops or finish harvesting in time and incur timeliness costs.

5.1.2 *Phytotechnology*

An alternative approach would be to use available information technologies to automate the processes to the point where they do not need a human operator. By removing the person from the immediate control of the system, it offers new opportunities but also creates new problems. Once the person is outside the control loop then the economies of scale that are applied to the larger manned tractors do not apply and alternative smaller smarter systems can be developed. Work rates (per day) can be kept high by working longer hours and using multiple machines.

Given that the scale of treatment areas have reduced over the last few years, due to the adoption of Precision Farming practices, from farm scale down to sub-field scale, this reduction could continue down to individual plant-scale treatments. This approach then becomes close to what people do when caring for plants in their own gardens, but is not usually realistic in a commercial environment. To make this approach practically viable, the machines must have enough intelligence embedded within them to carry out these tasks without anyone present. This concept called Phytotechnology was first described by Shibusawa et al. (2000).

By taking a systems approach to designing phytotechnology, consideration can be given to a system in terms of its action, interactions and implications. The result should be a new mechanization system that collectively deals with the crop's agronomic needs in a better way than is done now. Most people define agronomic processes in terms of how they are currently carried out and a break from this mentality, or paradigm shift, is needed to define the processes in terms of the fundamental plant needs. When the plant requirements are defined independently of the machine that carries out the processes, this improved specification can be used in conjunction with mechatronic principles to help design smarter, more efficient machines.

5.1.3 *Traditional Concepts*

A number of traditional concepts of plant production are now being reconsidered in the light of developing smarter machines.

During primary cultivation we invert the whole topsoil of a field with a plough to create a suitable seed bed. This is a generic operation that suits many circumstances but it uses a lot of energy. If this is turned around to consider the seed requirements, other options may become clearer. The seed requires contact with soil moisture to allow uptake of water to start emergence. Later on the root requires the same contact to uptake water and nutrients; the soil should have a structure that can hold the plant upright but also allow the roots to develop and the shoots to grow. If this same seed environment can be achieved by only mixing the soil within a few centimetres of the actual seed, then the rest of the soil does not need to be disturbed, as it can be well conditioned by natural soil flora and fauna when heavy machinery traffic is eliminated.

Another traditional concept is to grow crops in rows. It would seem that the only explanation as to why this is done is that it requires the simplest type of machines. Seeds are placed relatively densely along each row and we rely on crop plasticity to fill out the field surface. The problem is that, in principle, each plant requires equal access to light, air, water and nutrients, which are often spatially related. Intra-crop competition can be reduced by giving a more even or equal spacing, which in turn requires more accurate placement of seed by placing seeds in a more uniform pattern.

If the location of each seed is known and the position of each emerged crop plant is estimated, we can identify each plant by its spatial location. Improved information about plant characteristics allows improved management and decision making and allows a number of improved, more targeted operations that can improve the overall efficiency of growing the crop to be carried out.

The traditional arrangement of tractors supplying the motive power and humans supplying the intelligence, and using trailed or mounted equipment, can be reconsidered in the light of smarter machines. As the smarter implements are obviously more task-specific it may be better to have them self-propelled (having the disadvantage of low use over the year) or make the implement take control of the tractor. In this way the intelligence needed to carry out a specific task can be embedded within the appropriate machine, while the motive force (the tractor) could be generic.

At present machine efficiency (in terms of work rates) is being improved by increasing machine size, and improved efficacy can be achieved by adding driver assist technologies and limited machine intelligence such as auto-steer and variable rate applications.

5.2 Sustainability in Soil Engineering

5.2.1 Tillage Justification

The variation of soil level during the period after the primary tillage operation until crop sowing is shown in Fig. 5.1 (Koolen and Kuipers 1983). After tillage by ploughing, the soil is disturbed and an increase of its porosity increases its volume and its level. Several natural factors such as rainfall, or clay shrinkage and swelling, contribute to a re-compaction of the soil, but water freezing contributes to soil loosening. It is clear that the major soil compaction as shown by the reduction of its level should be attributed to the machinery movement on the soil. Seedbed preparation, sowing and other activities and mainly harvesting with very large and heavy machines cause soil compaction at a level that impedes root development, inhibiting plant growth and reducing yields.

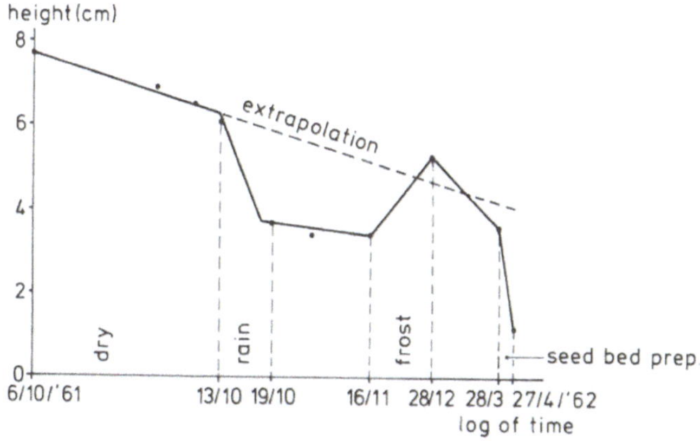

Fig. 5.1 Soil level changes during the season

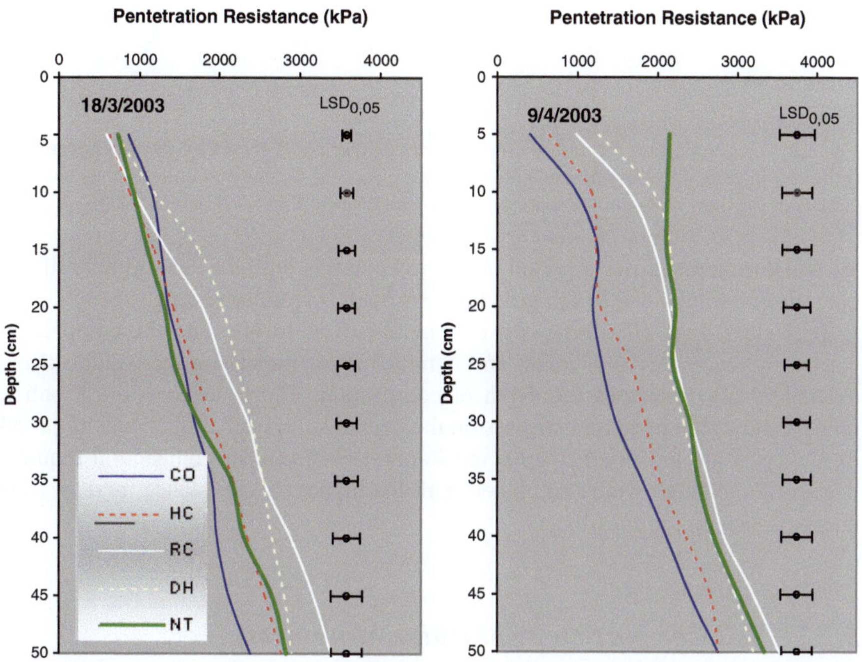

Fig. 5.2 Effect of the growing season traffic on soil compaction measured by soil penetration resistance for different soil tillage methods. CO is conventional tillage using plough at 25 cm depth, HC primary tillage using a heavy cultivator at 20 cm depth, RC primary and secondary tillage using a rotary digger at 12 cm depth, DH primary and secondary tillage using a disk harrow at 8 cm depth, NT is no-till plots (Cavalaris unpublished data)

The same effect is shown in Fig. 5.2 from an experiment in Greece (Cavalaris 2004, unpublished data). Soil penetration resistance was low after tillage but it was higher after seedbed preparation and planting. Figure 5.3 shows the compaction of

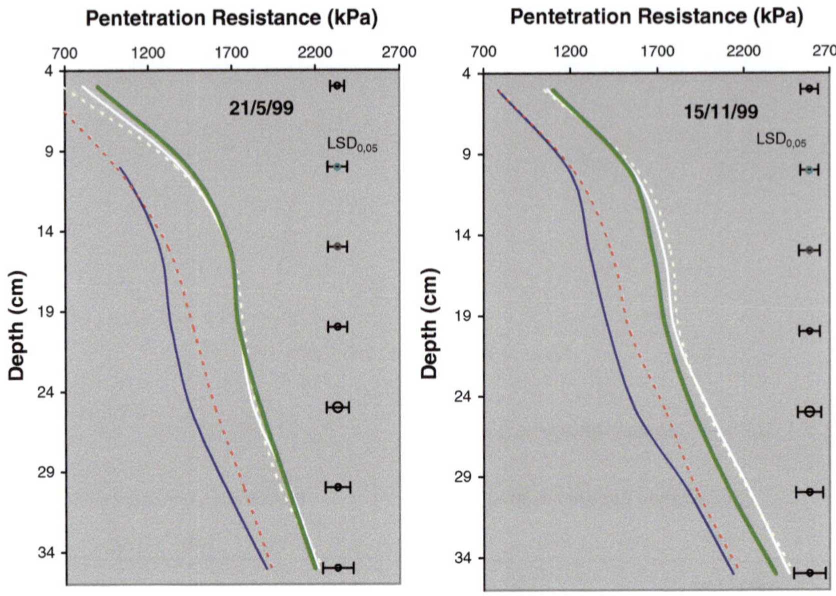

Fig. 5.3 Soil compaction increase during the growing period as shown by the soil penetration resistance increase in an experiment comparing the effects of different soil tillage practices. The colours of the lines represent the same tillage operations as in Fig. 5.2 (Cavlaris 2004)

the soil during the growing period (after crop planting until the next primary tillage operation) until the end of the growing season after crop harvesting.

It is clear that soil tillage is a continuous effort to alleviate the compaction caused by machinery movement and natural forces. Increasing the weight of the machinery also increases the depth of compaction. The Bousinesq and Frohlich models shows the increased stresses on the deeper soil layers as the loads imposed by the tyres are increased (Gupta and Raper 1994). Deeper compaction requires deeper tillage depth which combined with the higher soil resistance increases the energy consumption for tillage.

5.2.2 Soil Compaction by Machine Movements

Today's mechanized agriculture in the high labour-cost countries is based on large machines that offer size economies. The increased size of farm machinery contributes to a decreased cost of labour per unit of land. The ever-increasing size of tractors and harvesting equipment causes a lot of concerns. Several problems are caused such as soil compaction, difficulties in farm management, huge depreciation costs, etc. Several solutions are seen in the farm machinery construction sector to reduce compaction problems. The development of tracks to replace wheels is one direction which seems to reduce the problem. However, the size of machinery has a limit in the width of the vehicles running in public roads. This limit is 3 m and some of the large tractors with wide tyres exceed already this limit. Narrow tracks developed recently can reduce this problem but the problem of soil compaction remains. A recent research report from Cranfield University (Ansorge and Godwin 2007, 2008) indicated that soil compaction by a large wheel at a load of 10.5 t compacted the soil to a depth of 600 mm. This compaction cannot be alleviated by the usual cultivation and required deep soil loosening at a depth of 460 mm at a cost of about €8 ha^{-1}.

Soil compaction is of great concern in modern agriculture. Earlier research by Arvidsson (2001) and Arvidsson et al. (2001) indicated that the compaction caused by machinery with loads exceeding 15 tonnes per axle reached a depth beyond the usual tillage depth and the residual effect to the crops is apparent for four years. Chamen and Longstaff (1995) have proved that most tillage activities are made necessary by the soil compaction caused by the movement of machinery in the field. The Gantry system developed at the Silsoe Institute with a span of 12 m (i.e. 12 m without any machinery traffic) presented more than 30% energy saving for the field operations compared to conventional tillage with traffic.

5.2.3 Reduced and No-Till Problems

Several reported research projects have indicated that reduced or no till caused increased soil compaction and increased weed infestation that led to reduced yields (Cavalaris 2004; Cavalaris and Gemtos 2000). However, reduced tillage offers several advantages for soil fertility such as increased organic matter content, improved structure and water infiltration, protection from rain drop impact that decreases erosion, and reduced labour and energy requirements. If reduced size machinery was used, reduced or no tillage could be applied without the adverse effects of increased compaction.

5.2.4 Labour Cost and Mechanization

Field work cost is a combination of fixed costs (depreciation, capital cost, housing, and administration) and variable costs (driver, fuel and lubrication, repair and maintenance, and capital cost). Table 5.1 shows the cost for ploughing one ha with two New Holland tractors with and without a driver. The ploughing cost is slightly lower for the small tractor without the driver cost for the same use per year but it is lower for the larger unit when the driver is added. It should be noted that it is much easier for a small autonomous unit to work for many hours per day and achieve high yearly uses than the larger unit. Pedersen et al. (2005) conducted a study to compare the use of autonomous vehicles in comparison with conventional machinery for three scenarios: robotic weeding in high value crops (particularly

Table 5.1 Cost of ploughing for a 142 kW tractor (T7040) and 31.9 kW tractor (T3020) with and without a driver for different yearly use and for 15 years of economic life

	Working hours per year	T3020	T7040
Cost per ha	350	77.63	82.73
	500	71.22	75.16
	750	66.70	69.83
	1000	64.87	67.67
Adding driver cost at €10 h^{-1}			
Cost per ha	350	113.35	90.86
	500	106.94	83.29
	750	102.42	77.96
	1000	100.59	75.80

sugar beet), crop scouting in cereals, and grass cutting on golf courses. They concluded that in all three scenarios the autonomous vehicles were more economical than the conventional systems, while the high cost of RTK-GPS (real time kinematic global positioning system) and the low capacity of the autonomous vehicles were the main costs for autonomous systems. A question can be raised about the cost of the high accuracy RTK-GPS and the electronics of the autonomous vehicle systems. Even if the initial costs are high, experience up to now shows that electronics costs are decreasing over the years and the final costs will be low.

5.2.5 *Robotic Tractors Remove the Labour Cost Problem and Could Lead to an Alternative Mechanization System*

Blackmore et al. (2007), investigating the effects of the development of autonomous agricultural vehicles, suggested the use of a fleet of small tractors which would work day and night with small periods of resting for maintenance. Without a driver, the tractors will have a cost equal to the large tractors, with many advantages such as reduced soil compaction and the possibility to reduce soil tillage to a minimum. Additionally, modular farm mechanization can be developed. Small increases in area of the farm could easily be handled with an additional low-cost small tractor. This is rather difficult with existing large tractor mechanization. The development of robotic autonomous tractors seems to offer the opportunity to develop a new farm mechanization system profitable to the farmers and to the environment.

5.3 Systems Requirements for Robotics in Agriculture

A number of systems requirements have been identified as useful for enhancing the conceptual system of robotics in agriculture (Blackmore et al. 2004, 2007). The most predominant with influence on sustainability are light weight, small size, computational and energetic autonomy, machine intelligence, and external behaviour.

5.3.1 Light Weight

All agricultural machines should be as light as possible as this implies reduced soil compaction. Chamen et al. (1994) has identified that a 70% energy saving can be made in cultivation energy by moving from traditional trafficked systems (255 MJ ha^{-1}) to a non-trafficked system (79 MJ ha^{-1}). This was for shallow ploughing and did not include any deep loosening. From this we estimate that 80–90% of the energy going into traditional cultivation is there to repair the damage done by large vehicles. As the autonomous vehicles are inherently light, they should also require lower energy inputs and induce less soil compaction.

5.3.2 Small Autonomous Vehicles

The first autonomous vehicles should be small for a number of reasons. Firstly when a system failure occurs it is less likely that the resulting random action will be property- or life-threatening. Secondly, as there is no longer an operator on board then there is no need to carry that mass, nor the safety cab. Thirdly, the small machines could have the same work rates per day as manned ones if multiple machines were used for longer periods with the possibility of 24-hour operations. The smaller size and slower speeds also impliy higher precision, with the possibility to produce more self-propelled implements that are designed for a specific purpose such as physical weeding or crop scouting.

Multiple small machines are highly scaleable in that if higher work rates are needed then another machine could be purchased or borrowed. As these machines can be produced locally, the investment can be made incrementally each year rather than on a large tractor every few years.

It is envisaged that these machines could be made from existing agricultural and automotive parts that are already in production, which will significantly reduce the price of production. A current (2006) student project at the University of Thessaly in Greece is designing and evaluating a 4WD/4WS crop scouting robot being constructed by a skilled rural blacksmith using easily available parts.

5.3.3 Computational Energetic Autonomy

There are many levels of autonomy from automatic subsystems through to total autonomy, although this is very rare as some level of human intervention is usually required. Computational autonomy deals with the problem of how to

Gupta SC, Raper RL (1994) Predicting of soil compaction under vehicles. In: Soane BD, van Ouwerkerk C (eds) Compaction in Crop Production. Elsevier Science B.V., Netherland

Hannan WM, Burks FT (2004) Current Developments in Automated Citrus Harvesting. ASAE Paper No. 043087. St. Joseph, MI, ASAE.

Have H, Blackmore S, Keller B, Fountas S, Nielsen H, Theilby F (2002) Autonomous weeder for Christmas tree plantations – a feasibility study. In Proc. EurAgEng02: European Conference of Agricultural Engineering, June 30-July 4, Budapest, Hungary, Paper No. 02-AE-023: EurAgEng.

Hollingum J (1999) Robots in agriculture. Ind Robot 26(6):438–445

Ieropoulos I, Greenman J, Melhuish C (2003) Imitating metabolism: Energy autonomy in biologically inspired robots. In Proc. AISB '03 Second international symposium on imitation in animals and artifacts, 191–104. Aberystwyth, Wales, 7–11 April.

Kataoka T, Okamoto H, Hata S (2001) Automatic detecting system of apple harvest season for robotic apple h arvesting. ASAE Paper No: 013132. St. Joseph MI, ASAE.

Kelly I, Holland O, Melhuish C (2000) SlugBot: a robotic predator in the natural world. 5th International symposium on artificial life and robotics, 470–475. Oita, Japan, 26–28 January.

Kondo N, Ting KC (1998) Robotics for bioproduction systems. ASAE Publication, St. Joseph, MI

Kondo N, Ninomiya K, Hayashi S, Ohta T, Kubota K (2005) A new challenge of robot for harvesting strawberry grown on table top culture. ASAE Paper No: 043083. St. Joseph MI, ASAE.

Koolen AJ, Kuipers H (1983) Agricultural soil mechanics. Springer, Berlin, p 241

Pedersen SM, Fountas S, Have H, Blackmore BS (2005) Agricultural robots – system analysis and economic feasibility. Precis Agr 7(4):295–308

Pilarski T, Happold M, Pangels H, Ollis M, Fitzpatrick K, Stentz A (2002) The Demeter system for automated harvesting. Auton Robot 13(1):9–20

Shibusawa S, Anom IM, Sata H, Sasao A, Hirako S, Otomo A, Blackmore S (2000) On-line real-time soil spectrophotometer. In Proc. 5th International Conference on Precision Agriculture. Robert PC, Rust RH and Larson WE (Eds.). ASA, CSSA, SSSA & ASAE.

Turing A (1950) Computing machniery and intelligence. Mind 59:433–460

Wilson JN (2000) Guidance of agricultural vehicles – a historical perspective. Comput Electron Agr 25:1–9

Chapter 6
Soil Density Increases resulting from Alternative Tire and Rubber Track Configurations in Laboratory and Field Conditions

Dirk Ansorge and Richard John Godwin

6.1 Introduction

The continuous demand for more efficient agricultural machinery provides the challenge of contradictiory demands for agricultural machinery manufacturers. For economic reasons the machines have to increase in size and hence in weight. However, the soil compaction caused by the undercarriage gear carrying these machines should not exceed the bearing capacity of the soil (Raper 2005).

There are several alternatives to meet this challenge:

- With increasing machine size, individual axle loads can be maintained when the increase in machine weight is accounted for by additional axles. However, these extra axles and tires would contribute a significant extra weight to the machine and occupy space which alternatively could be used for the processing technology within the machine.
- Another option would be the use of light composite materials and light metal alloys. However, these materials are expensive and thus their abundant use is rather theoretical in harvest machinery.
- Alternatively large volume tires could be used providing low ground contact pressure. But there are physical limitations to this approach. The ground contact patch would have to increase. This could be realized by larger diameter or wider tires. Wider tires face difficulties because of the overall width of the machine, at least under European conditions. Larger diameter tires face the challenge of being able to fit into a given space within the machine. For example imagine a combine harvester with a large diameter tire on the front axle – physically, the limit would be reached when the driver was not able to easily exit the cabin. Moreover low

D. Ansorge (✉)
CLAAS Selbstfahrende Erntemaschinen GmbH, Münsterstr. 33, 33428, Harsewinkel, Germany
e-mail: Dirk.ansorge@claas.com
R.J. Godwin
Cranfield University, MK43 0AL, Cranfield, Bedfordshire, UK

A.P. Dedousis and T. Bartzanas (eds.), *Soil Engineering*, Soil Biology 20,
DOI 10.1007/978-3-642-03681-1_6, © Springer-Verlag Berlin Heidelberg 2010

inflation pressure tires can only protect the soil to a certain extent, as soil compaction at depth is mostly determined by the overall tire load (Soehne 1953).
- A further option providing low ground pressure is track concepts. Track concepts can maintain a low ground pressure while being lower in size and narrower than tires. Recent studies conducted by the authors have focused on the soil compaction caused by rubber tracks in comparison to commercially available tires, as the publicly available literature provides little suitable information.

The following literature review will focus on the contradictory results published in the public domain on the effect of track and tire undercarriage systems with respect to soil density increase.

6.2 Literature Review

Erbach (1994) pointed out that tracks in general were better at limiting soil compaction than tires. Yet tracks, both rubber- and steel-belted ones, could have detrimental effects on soil because the contact pressure, although lower in magnitude than for a tire, was applied for a longer time period(Culshaw 1986; Erbach 1994). Moreover, these authors pointed out that a non-uniform pressure distribution may result from belts with inadequate tension and idler configuration. Last but not least, machine vibrations originating from the engine and other vibrating/moving machine parts were efficiently transferred into the soil.

The benefit of steel tracks with respect to soil compaction were found in studies published by Kinney et al. (1992), Erbach et al. (1991), Erbach et al. (1988), Janzen et al. (1985), Taylor and Burt (1975), Soane (1973), and Reaves and Cooper (1960). In contrast to the previous publications, Burger et al. (1985) and Burger et al. (1983) could not detect differences between steel-tracked and rubber-tired tractors; the results were influenced not by contact pressure, but other machine-related factors.

An investigation into the compaction behavior of steel tracks, rubber tracks, and wheels by Brown et al. (1992) showed that rubber tracks produced an intermediate amount of soil compaction, significantly different neither from tires nor from steel tracks. The authors attributed these findings to the less rigid belt of rubber tracks and the idler configuration, which could produce an uneven pressure distribution below the belt on soft surfaces. Unequal pressure distribution underneath rubber tracks was also reported from studies by Tijink (1994), Keller et al. (2002), and Weissbach (2003).

Alakukku et al. (2003) summarized papers reporting disadvantages (Blunden et al. 1994) or advantages (Bashford et al. 1988; Rusanov 1991) of tracks on soil compaction.

The variety of results emphasized the importance of the design of the track frame carrying the idlers and tensioning the rubber belt in order to apply contact pressure uniformly onto the soil. Only if this was achieved would rubber tracks provide an undercarriage gear supporting high loads with a small resulting soil compaction. This was less crucial for steel tracks due to the higher inherent strength of the steel belt.

All studies reported on previously were conducted in field situations. Field studies always bear the risk of environmental influences and varying field conditions in moisture content, organic matter content and soil textures. Hence, a large number of samples and replications were compulsory in order to derive a reliable answer.

Because of the above situation, a study under controlled laboratory conditions was conducted at Cranfield University, Silsoe. The aim was to derive repeatable results under controlled conditions to provide a qualitative answer for the potential of a rubber track system to minimize soil compaction. Subsequently, a field study was conducted to confirm the overall conclusions derived in the laboratory. The results were published in full in Ansorge and Godwin (2007, 2008). This chapter summarizes the findings of both papers with respect to the measured soil density increase. Additionally, penetrometer resistence and gravimetric dry bulk density were measured.

6.3 Material and Methods

The study was conducted on a sandy loam soil with a particle size distribution of 17% clay, 17% silt, and 66% sand. The organic matter content was 4.1%. The soil was prepared in layers of 50 mm to a weak uniform density of 1.38 g cm^{-3} with the soil processor of the soil bin laboratory whereby a uniform gravimetric moisture content of 10% was maintained throughout the study. The low density of 1.38 g cm^{-3} was particularly chosen to amplify the differences between the single treatments and to reflect weaker soil conditions when tracks might provide an advantage. During the preparation, talcum powder lines were embedded into the soil in order to provide traceability of soil movement caused by the passage of undercarriage gear. Full details of the preparation of the soil bin and the talcum powder technique including the determination of the resulting soil displacement and soil density increase are given in Ansorge and Godwin (2007). For the entire study, a large range of tires were used. Table 6.1 contains the specifications for the tires and tracks reviewed in this article.

Table 6.1 Tire and track specifications

Implement	Specifications	Width (mm)	Load (t)	Inflation pressure (bar)	Abbreviation
Rubber track	CLAAS terra trac	635	12	–	T12
Rubber track	CLAAS terra trac	635	10.5	–	T10.5
Tire	900/65 R32	900	10.5	1.9	900/10.5/1.9
Tire	800/65 R32	800	10.5	2.5	800/10.5/2.5
Tire	800/65 R32	800	10.5	1.25	800/10.5/1.25
Tire	710/45–26.5	710	4.5	1.0	700/4.5/1.0
Tire	23.1–26		4.0	1.2	23/4/1.2
Tire	11.5/80–15.3		1.5	2.0	11/1.5/2.0

Fig. 6.1 Single wheel/track test apparatus with a tire (*left-hand side*) and a track (*right-hand side*)

The undercarriage gear simulating the running gear of harvest machinery was placed into a single wheel/track tester. The test apparatus shown in Fig. 6.1 was specifically built for the study and could transfer up to 15 t onto a self-propelled tire or track. The tire or track propelled itself through the soil bin over a length of approximately 10 m at a speed of 1 m s^{-1}, which represented a custom speed for soil compaction analysis. Unpublished investigations by Ansorge showed that the resulting compaction was approximately 20% higher at 1 m s^{-1} compared to higher speeds in the range of 1.5–2 m s^{-1}. This was acceptable since the study aimed at a qualitative determination of the soil displacement and all undercarriage systems were tested at the same speed.

Multiple passages were simulated by additional passages of the single wheel/track tester. Small non-driven tires were pulled through the soil bin in a separate single wheel tester by the soil processor.

6.4 Results

6.4.1 Single Axle Comparison

When considering the resulting soil displacement caused by single passages of rubber tracks and tires in these volatile soil conditions, chosen to amplify the differences between the individual treatments, the results showed a clear benefit of the tracks with respect to minimizing soil displacement and hence reducing the resulting soil compaction.

The results in Fig. 6.2 show that a rubber track unit at 10.5 t (T10.5) and 12 t (T12), respectively, causes similar soil displacement as an implement tire at a load

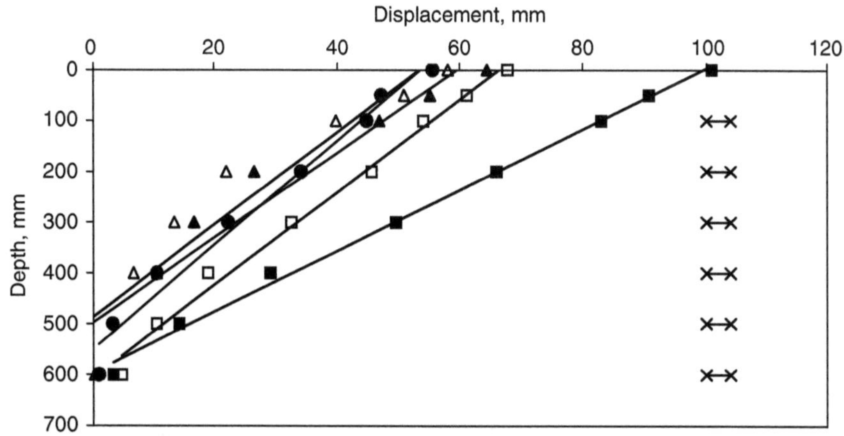

Fig. 6.2 Soil displacement over depth for a rubber track at 10.5 t (T10.5) (*open triangle*), and at 12 t (T12) (*filled triangle*); an 800 mm wide harvester tire at recommended (800/10.5/2.5) (*filled square*), and half recommended (800/10.5/1.25) (*open square*) inflation pressure; and a implement tire (700/4.5/1.0) (*filled circle*), including the least significant difference bars (*times symbol*) (Following Ansorge and Godwin 2007)

of 4.5 t with an inflation pressure of 1 bar and a width of 700 mm (700/4.5/1.0). In fact, this showed that the rubber track created similar soil displacement, and hence compaction, as a tire carrying approximately one third of the load. The two track loads were chosen to represent both the scientific comparison, whereby the rubber track and tire ran at the same load (T10.5), and the practical comparison (T12), whereby the track ran at a 1.5 t higher load, accounting for the higher initial weight of the track unit compared to the tire.

The large harvester tire at a load of 10.5 t and an inflation pressure of 2.5 bar at a width of 800 mm (800/10.5/2.5) caused significantly more soil displacement. If for scientific purposes the inflation pressure of that tire was reduced to 1.25 bar at the same load (800/10.5/1.25), rut depth reduced significantly. However, at a depth of 500 mm, the displacement approached that of a tire at the recommended higher inflation pressure.

This result confirmed the theory of Soehne (1953) stating that the inflation pressure determined the soil compaction and soil movement at the surface. However, overall load determined the effect at depth.

In contrast to the tires, both tracks (T10.5 and T12) maintain their benefit over the entire depth and approach zero displacement at 500 mm depth, whereas the tires maintained a residual displacement at 600 mm depth. Therefore the theory that contact pressure determined compaction at the surface and load at depth was not valid for this rubber track system. This finding supports the potential of rubber tracks at minimizing soil density increase throughout the soil profile.

Similar results were gained on stratified soil conditions resembling field soil conditions (Ansorge and Godwin 2007). Additionally, the results published by

Ansorge and Godwin (2007) contain a comparison of additional widths and inflation pressures of tires in a range of ~1.2–2.0 m in diameter.

6.4.2 Entire Machine Configurations

Following the investigation of soil compaction caused by single axles in Ansorge and Godwin (2007), large modern combine harvester axle configurations were simulated by Ansorge and Godwin (2008) with total machine weights of 30 t (21 t on front axle and 9 t on rear axle) and 33 t (24 t on front axle and 9 t on rear axle) for wheeled and half-tracked combine harvesters, respectively. As Fig. 6.3 shows, the soil displacement caused by the simulated combine harvester equipped with half tracks was significantly smaller than that of a wheeled one.

The soil displacement caused by the rubber-tracked combine harvester was similar to that of a smaller wheeled machine at a total machine weight of 11 t (8 t on front axle and 3 t on rear axle) in which the front axle tire width was chosen to represent the rubber belt width. This was the intermediate tire size available for a combine harvester of that size and weight. The rubber track unit followed by a smaller tire caused similar compaction to a wheeled machine configuration carrying approximately one third of the weight.

Ansorge and Godwin (2008) showed that the size of the rear tire, i.e. the second tire passing the soil, had a larger influence after tires than after tracks, although tires created more compaction in the first passage. This was due to the lateral soil movement caused by the rubber track within the surface 150 mm of the soil. The track created a path for the following tire which was able to support the load of that tire. Fig. 6.4 shows schematically the position of the sand columns indicating the

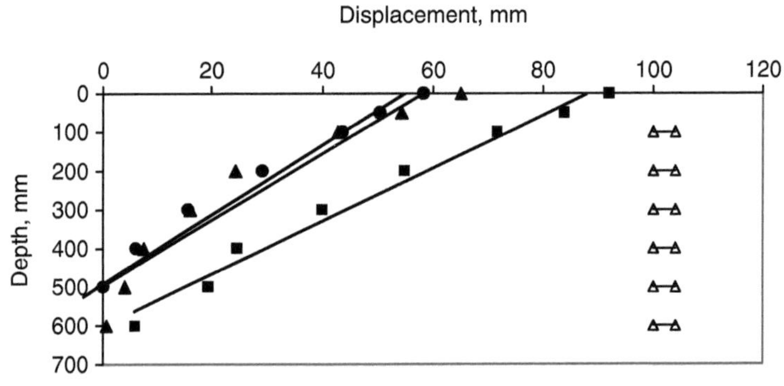

Fig. 6.3 Soil displacement over depth for a rubber track at 12 t (T12) followed by a 700/4.5/1.0 tire (*closed triangle*); an 900 mm wide harvester tire at recommended inflation pressure (900/10.5/1.9) followed by a smaller rear axle tire (700/4.5/1.0) (*filled square*), and a light machine combination with 23/4/1.2 on the front axle and 11/1.5/2.0 on the rear axle (*closed circle*), including the least significant difference bars (*times symbol*) (Following Ansorge and Godwin 2008)

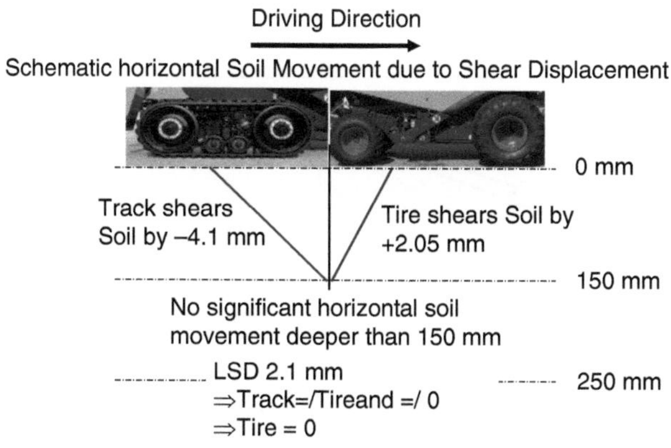

Fig. 6.4 Longitudinal soil movement caused by a self-propelled track and tire carrying its load and without further draft force applied

lateral soil movement after the passage of a tire and after the passage of a track. It was apparent that the soil after the passage of a track moved backwards whereas after that of a tire the soil moved forward due to the bow wave of the tire. This lateral soil movement was limited to the surface 150 mm in these soft uniform soil conditions for both tire and track and was expected not to exceed that depth in the field.

Subsequently, the soil bin laboratory results were validated with full-size combine harvesters in field conditions on a sandy loam and a clay soil. To this end, a technique was developed to measure soil displacement in the field by Ansorge and Godwin (2009). The authors utilized treble fishing hooks, with a fishing line of known length attached to them. These were pushed into the soil to an approximate depth with a thin rod. Subsequently, the length of the fishing line remaining above the ground was measured to a reference surface. After the passage of the machine, the hook was carefully excavated following the line and before it was removed from the soil, the length of the fishing line above the reference surface was measured again. From the difference in the two lengths, the vertical soil movement could be accurately determined as shown in Fig. 6.5, using comparison tests in the laboratory where the fishing hooks were compared to the talcum powder lines.

Fig. 6.6 shows that the field results validated the overall conclusions found in the laboratory. As apparent, the scatter of the data was bigger due to the variable field conditions and movements of the datum surface due to heavy precipitation after embedding the fishing hooks. The soil displacement of the machine equipped with a half rubber track system at a total load of 24 t approaches zero at a depth of 375 mm. In contrast the soil displacement caused by the wheeled machine at a total load of 21 t approaches zero at approximately 500 mm depth. The linear regression lines do not follow this behavior at depth exactly, but overall give a good indication of the different soil displacement. Total machine weights are less than the ones simulated in the laboratory due to empty grain tanks. This was accepted, as the aim of the

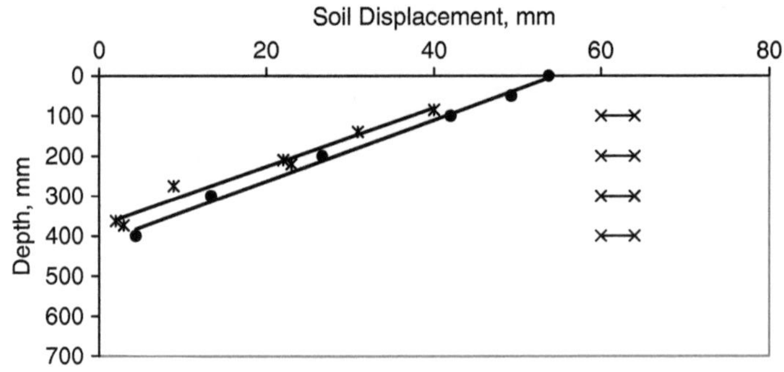

Fig. 6.5 Soil displacement measured with talcum powder lines (*closed circle*), and with the fish-hook method (*asterisk*), in soil bin conditions for 23/4/1.2 on the front axle and 11/1.5/2.0 on the rear axle including the least significant difference bars (*times symbol*)

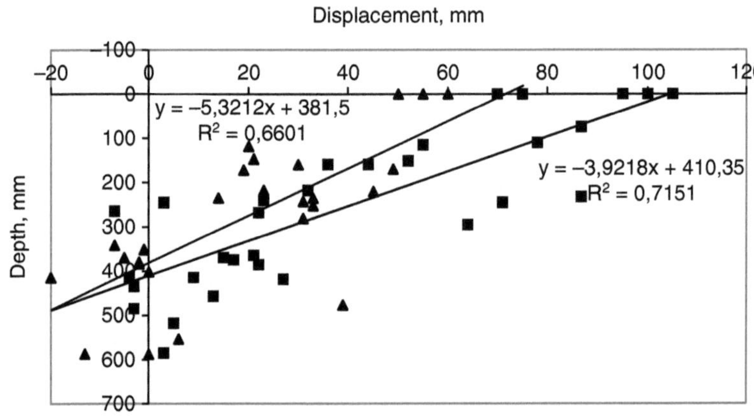

Fig. 6.6 Soil displacement over depth for a combine harvester with the half-track system at 24 t overall load (*closed triangle*), and for a wheeled machine at 21 t overall load at recommended inflation pressure (*filled square*), including the least significant difference bars (*times symbol*)

study was a comparison of the undercarriage systems with identical conditions. The weight difference was due to the additional weight of the rubber track system.

6.5 Conclusions

Ansorge and Godwin (2007, 2008) conducted the first in-depth investigation into the effect of different undercarriage systems with respect to the resulting soil displacement. This work contributed significantly to the discussion on the potential of rubber tracks at limiting soil density increase.

It can be concluded that the rubber track system provided a means to significantly reduce soil displacement and hence soil compaction. In absolute numbers, soil displacement for tracks was maintained approximately in the range of soil displacement caused by a wheeled undercarriage system with about one third of the weight.

Lateral soil movement was facing backwards after the passage of a track and facing forwards after the passage of a tire, and limited for both to within the common working depth. The lateral soil movement caused by the track created a pathway which was able to carry the load of the subsequent tire passage.

The talcum powder lines provide the possibility for an accurate determination of soil displacement in the vertical and horizontal direction caused by the passage of tires and tracks in soil bin conditions. For the in-field method, the use of fishing hooks provided a reliable answer when compared to the talcum powder method used in the laboratory, providing care was taken to provide a reliable surface datum.

Acknowledgements The authors want to thank the CLAAS Company, Harsewinkel, Germany, for its support and Gordon Spoor for useful suggestions concerning the work. Thanks must go to Prof. Kutzbach from the University of Hohenheim for enabling Dirk Ansorge to participate in the Double Degree Program under which these parts of the study were conducted. The help from Prof. Piepho from the University of Hohenheim in analyzing the data statistically was very valuable.

References

Ansorge D, Godwin RJ (2007) The effect of tires and a rubber track at high axle loads on soil compaction, part 1: single axle studies. Biosyst Eng 98(1):115–126

Ansorge D, Godwin RJ (2008) The effect of tires and a rubber track at high axle loads on soil compaction, part 2: multi-axle machine studies. Biosyst Eng 99(3):338–347

Ansorge D, Godwin RJ (2009) An in-situ determination of virgin compression line parameters for predicting soil displacement resulting from agricultural tire passes. J Terramechanics 46(5):251–258

Alakukku L, Weisskopf P, Chamen WCT, Tijink FGJ, van den Linden JP, Pires S, Sommer C, Spoor G (2003) Prevention strategies for field traffic-induced subsoil compaction: a review. Part 1. Machine/soil interactions. Soil Tillage Res 73(1–2):145–160

Bashford LL, Jones AJ, Mielke LN (1988) Comparison of bulk density beneath a belt track and a tire. Appl Eng Agric 4(1):122–125

Blunden BG, McBride RA, Daniel H, Blackwell PS (1994) Compaction on an earthy sand by rubber tracked and tired vehicles. Aust J Soil Res 32(6):1095–1108

Brown HJ, Cruse RM, Erbach DC, Melvin SW (1992) Tractive device effects on soil physical properties. Soil Tillage Res 22(1–2):41–53

Burger JA, Perumpral JV, Kreh RE, Torbert JL, Minaei S (1983) The effect of track and rubber-tired vehicles on soil compaction. ASAE Paper No. 83–1621

Burger JA, Perumpral JV, Kreh RE, Torbert JL, Minaei S (1985) Impact of tracked and rubber-tired tractors on a forest soil. Trans ASAE 28(2):369–373

Culshaw D (1986) Rubber tracks for agriculture. Div. Note DN 1345, Natl Inst Agric Engng, Silsoe, UK

Erbach DC, Melvin SW, Cruse RM, (1988) Effects of tractor tracks during secondary tillage on corn production. ASAE Paper No. 88–1614

Erbach DC, Melvin SW, Cruse RM (1991) Low ground pressure equipment fleets for crop production. ASAE Paper No. 91–1517
Erbach DC (1994) Benefits of tracked vehicles in crop production. In: Soane BD, van Ouwerkerk C (eds) Developments in agricultural engineering 11. Soil compaction in crop production. Elsevier, Amsterdam, pp 501–520
Janzen DC, Hefner RE, Erbach DC (1985) Soil and corn response to track and wheel compaction. In: Proceedings International Conference on Soil Dynamics, National Soil Dynamics Laboratory, Auburn, AL, USA
Keller T, Trautner A, Arvidsson J (2002) Stress distribution and soil displacement under a rubber tracked and a wheeled tractor during ploughing, both on land and within furrows. Soil Tillage Res 68(1):39–47
Kinney GR, Erbach DC, Bern CJ (1992) Soil strain under three tractor configurations. Trans ASAE 35(4):1135–1139
Raper RL (2005) Agricultural traffic impacts on soil. J Terramechanics 42(3–4):259–280
Reaves CA, Cooper AW (1960) Stress distribution in soils under tractor loads. Agric Eng 41(1):20–21 31
Rusanov VA (1991) Effects of wheel and track traffic on the soil and crop growth and yield. Soil Tillage Res 19(2–3):131–143
Soane BD (1973) Techniques for measuring changes in the packing state and cone resistance of soil after the passage of wheels and tracks. J Soil Sci 24(3):311–323
Soehne W (1953) Druckverteilung im Boden und Bodenverformung unter Schleppereifen. Grundlagen Landtechnik 5:49–63
Tijink FGJ (1994) Quantification of vehicle running gear. In: Soane BD, van Ouwerkerk C (eds) Developments in agricultural engineering, 11. Soil compaction in crop production. Elsevier, Amsterdam, pp 391–415
Taylor JH, Burt EC (1975) Track and tire performance in agriculture. Trans ASAE 18(1):3–6
Weissbach M (2003) Landtechnische Untersuchungen zur Wirkung bodenschonender Fahrwerke an Schleppern und Arbeitsmaschinen mit verschiedenen Radlasten [Agricultural engineering research with of soil protecting drive systems from tractors and self proplled agricultural machinery.] Habilitationsschrift, University of Kiel. Logos Verlag, Berlin

Chapter 7
Effects of Heavy Agricultural Machines for Sugar Beet Harvesting on Physical Soil Properties

Rupert Geischeder, Markus Demmel, and Robert Brandhuber

7.1 Introduction

Agricultural land use is facing an increasing risk of heavy machines causing soil compaction which might negatively impact soil functions (Chamen et al. 2003; Håkansson 2005). Self-propelled six-row bunker hopper sugar beet harvesters feature total weights up to 60 metric tons or more and are the heaviest agricultural machines, followed by self-propelled combine harvesters and self-propelled slurry tankers. To reduce the risk of soil compaction, alternative undercarriage concepts with multiple axles, crab steering and high volume wheels or with rubber belt tracks have been developed (Thangavadivelu et al. 1992; Tijink and van der Linden 2000). Only a few investigations of the stress on the soil resulting from traffic of heavy agricultural machinery with innovative undercarriage concepts have been reported (Keller and Arvidsson 2004; Gysi 2001; Schäfer-Landefeld et al. 2004) and little information is available on the consequences to physical soil properties. Vitally important is the question of whether rubber belt tracks or additional axles resulting in larger contact areas, but also with additional or longer lasting pressure impulses, can serve as soil protecting equipment for heavy agricultural machines.

By means of a field trial, the effects of typical trafficking and load situations caused by self-propelled six-row bunker hopper sugar beet harvesters on the soil structure beneath the topsoil were investigated.

The following questions were addressed:

R. Geischeder (✉)
HARAIN Engineering GmbH, Oberharthausener Str. 14, 94333, Geiselhoering, Germany
e-mail: rupert.geischeder@harain.com

M. Demmel
Institute for Agricultural Engineering and Animal Husbandry, Bavarian State Research Center for Agriculture, Voettinger Str. 36, 85354, Freising, Germany

R. Brandhuber
Institute for Agroecology, Organic Farming and Soil Protection, Bavarian State Research Center for Agriculture, Voettinger Str. 38, 85354, Freising, Germany

A.P. Dedousis and T. Bartzanas (eds.), *Soil Engineering*, Soil Biology 20,
DOI 10.1007/978-3-642-03681-1_7, © Springer-Verlag Berlin Heidelberg 2010

- Can the large contact area of rubber belt undercarriages reduce the risk of subsoil compaction compared to undercarriages with tires?
- Can the risk of subsoil compaction be reduced by distributing the load to more than four wheels in conjunction with reduced wheel loads?
- Does soil stress with only little deformation repeated year by year cause significant degradation of the soil structure at the end of three years?

7.2 Material and Methods

7.2.1 Field Experiment

A field of 3 ha with minor preload by agricultural field traffic in southern Germany, Lower Bavaria, was chosen for the field trial. Mean annual temperature and annual precipitation are 8°C and 0.70–0.75 m respectively. Soils are classified as luvisols, cambisols and leptosols developed from loess sediments. Mean soil particle size distribution (<2 μm/2–63 μm/63–2000 μm) is 0.21/0.65/0.14 $g\,g^{-1}$, characterizing a silty loam.

Six different trafficking situations caused by three self-propelled six-row bunker hopper sugar beet harvesters with distinct undercarriage systems were selected as treatments for further investigations:

1. Single pass with a large tire and reduced wheel load (bunker hopper nearly empty)
2. Single pass with a large tire and maximum machine load (filled bunker hopper)
3. Single pass with a rubber belt undercarriage and maximum machine load (filled bunker hopper)
4. Double pass with large tires and maximum machine load (filled bunker hopper)
5. Triple pass with large tires and maximum machine load (filled bunker hopper)
6. Double pass with rubber belt track plus large tire and maximum machine load (filled bunker hopper).

The treatments characterize typical trafficking patterns of self-propelled six-row bunker hopper sugar beet harvesters. Two- and three-axle machines work with crab steering and cause single, double and also triple passes. The machine with the combination of rubber belt undercarriage in front and wheels in the rear causes single passes of the belts and the wheels and a double path of both (Fig. 7.1).

Tire inflation pressure was adjusted as low as possible accordingly to the actual load determined in the field, to field conditions and to tire manufacturer's manuals. A complex experimental design was developed and realized using precise positioning by a Real Time Kinematic Differential Positioning System (RTK DGPS). During the investigation period (2004–2006) every year in autumn the three harvesters drove on identical tracks over the field. Alternately, one half of the field was planted with sugar beets, the other with winter wheat which was harvested in summer. In the field trial the machines actually harvested the sugar beets and drove over the wheat stubbles simulating sugar beet harvesting. Thereby crop rotation and controlled field traffic was guaranteed year by year. The field was not irrigated.

Fig. 7.1 Self-propelled six-row bunker hopper sugar beet harvesters used in the investigation, undercarriages and pattern of trafficking

7.2.2 Measurement Program

The first trafficking took place in autumn 2004. All machine parameters (axle- and wheel loads, tire inflation pressure, ground contact area) were determined as well as the soil stress during passes using hose-type pressure transducers (Bolling 1986). For investigating the physical soil properties, each year 1080 undisturbed soil samples were collected from two depths (0.28–0.33 and 0.38–0.43 m, 540 samples before and 540 samples after pass). At least 9 randomized repetitions per pass situation (treatment), each represented by five soil cores in each depth, were available for statistical analysis. The area for soil sampling before and after pass with the harvesters was 45 m wide, covered one-third of the field each year and lay in the middle of the field in 2004. In autumn 2005 and 2006, soil physical investigations were repeated the same way as in 2004 with the difference that the 45 m wide soil sampling areas were located in the south (2005) and the north (2006) of the area of 2004.

7.3 Results

7.3.1 Mechanical Soil Stress – Parameters of Trafficking Situations

The technical parameters of the six different trafficking situations (treatments) 2004–2006 are shown in Table 7.1. The maximum tire or rubber belt track loads were up to 140 kN, with partly filled bunker hopper 90 kN. The lowest mean ground contact pressure values were 94 kPa with the rubber belt undercarriage, the highest varying between 140 and 150 kPa with single and multiple passes, all with maximum machine load (filled bunker hopper).

Table 7.1 Mechanical soil stress: parameters of trafficking situations (means 2004–2006)

Treatment	Dimension of tracks and tires	Load per wheel or track (kN)	Tire inflation pressure (kPa)	Ground contact area (m^2)	Mean ground contact pressure[a] (kPa)
(1) Large tire	1050/50R32	91	130	0.92	100
(2) Large tire	1050/50R32	131	225	0.94	141
(3) Rubber belt track	0.89 × 2.0 m	147	–	161	94
(4) Double pass large tire	900/55R32	88	215	0.79	146
	1050/50R32	115	200	0.73	146
(5) Triple pass large tires	900/55R32	60	215	0.64	94
	1050/50R32	115	215	0.79	146
	1050/50R25	88	200	0.73	122
(6) Double pass track and tire	0.89 × 2.0 m	147	–	1.61	94
	900/60R32	101	165	0.88	118

[a]Ratio of ground contact area and load

Table 7.2 Mean volumetric water content in upper and second subsoil at the field trial, 2004–2006

Year	Upper subsoil (0.28–0.33 m depth)		Second subsoil (0.38–0.43 m depth)	
	Volumetric moisture content [m^3 100 m^{-3}]	Measured field capacity[a] [%]	Volumetric moisture content [m^3 100 m^{-3}]	Measured field capacity[a] [%]
2004	31.1	98	28.7	88
2005	29.2	89	27.1	81
2006	25.2	76	24.2	73

[a]Defined as the volumetric water content at a water tension of 6.2 kPa

7.3.2 Soil Moisture at Trafficking

At the date of trafficking in autumn 2004, 2005 and 2006 the upper subsoil (0.28–0.33 m) and the second subsoil (0.38–0.44 m) were characterized by quite different moisture contents based on the climate of the different years (Table 7.2). While in 2004 upper subsoil and second subsoil were relatively wet, in 2005 only the second subsoil and in 2006 neither the upper subsoil nor the second subsoil can be characterized as wet.

Because of the slowly penetrating moisture frontier in all three years the second subsoil was drier than the upper subsoil. The locally very dense sampling showed a high heterogeneity of the soil moisture especially in the second subsoil

7.3.3 Soil Stress Measurements

Only in 2004, soil stress measurement with hose-type pressure transducers at a depth of 0.45 m showed pressure peaks up to 150 kPa under the large tires

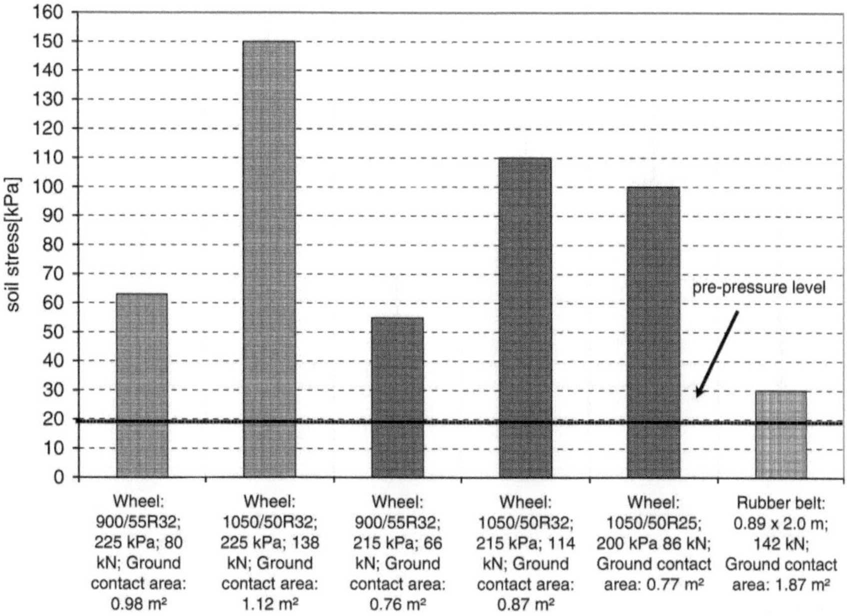

Fig. 7.2 Soil stress under the different undercarriages at a depth of 0.45 m measured with hose-type transducers (pre-pressure of 20 kPa)

depending on the wheel load and the tire size and up to 30 kPa under the rubber belt undercarriage (Fig. 7.2 and Table 7.3).

The values for the rubber belt in the "soil stress" column of Fig. 7.2 represents the mean of the pressure peaks under the four rubber belt axles.

Therefore, in another investigation in spring 2006, soil pressure data to characterize soil stress were determined very intensively. The results are comparable (Geischeder et al. 2008). Under rubber belt undercarriages soil pressure values decline faster with increasing depth than under tires (Ansorge and Godwin 2007).

7.3.4 Changes in Physical Soil Properties

7.3.4.1 Short-term Effects

The physical soil properties total porosity, coarse porosity (drained pore volume at field capacity) and pneumatic conductivity determined before and after trafficking show similar results. A statistical differentiation of the treatments is not significant using the parameter pneumatic conductivity because of the large variation in values, but is possible using total porosity and coarse porosity.

Figures 7.3 and 7.4 show the coarse porosity, which represents the part of the large pores (fast-moving leachate), one of the specific texture values, which

Table 7.3 Soil stress determined using hose-type pressure transducers at a depth of 0.45 m, 2 November 2004

Treatment	Dimension of tracks and tires	Load per wheel or track (kN)	Tire inflation pressure (kPa)	Ground contact area (m^2)	Soil stress / maximum soil pressure at a depth of 0.45 m (kPa)
(3) Single pass rubber belt track	0.89 × 2.0 m	140	–	1.87	30
(4) Double pass large tires	900/55R32	80	230	0.98	63
	1050/50R32	140	230	1.12	150
(5) Triple pass large tires	900/55R32	70	215	0.76	57
	1050/50R32	110	215	0.87	110
	1050/50R25	90	200	0.77	100

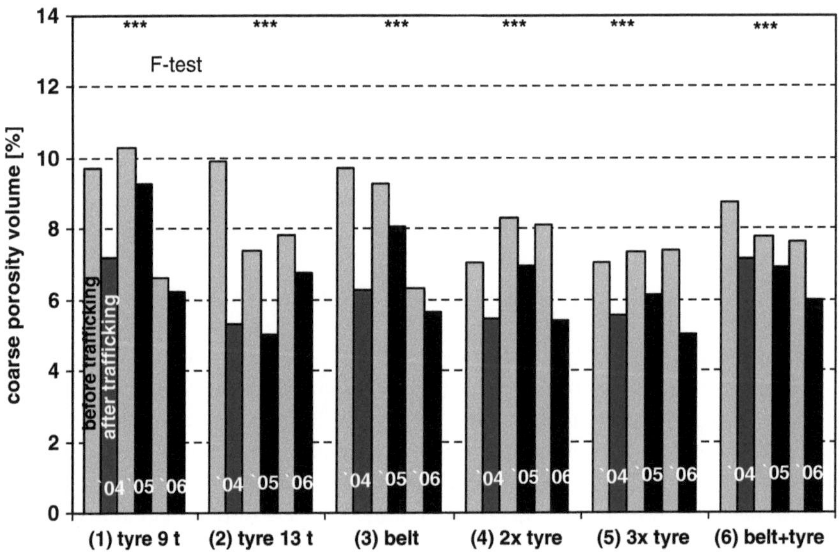

Fig. 7.3 Effects of different level of soil stress in the upper subsoil (0.28–0.33 m): coarse porosity before (*gray columns*) and after (*black columns*) trafficking in 2004, 2005 and 2006

according to Lebert et al. (2004) delivers information about soil damaging compaction. A critical level for the subsoil is 5% coarse porosity. The values which are represented by the columns are the average of 9 sampling locations in the field (each 5 soil cores, 45 soil cores per column). The significance of the effects of the treatments was determined by analysis of variance using an F-test. Shown in the figures is the significance of the total effect over 3 years/3 times trafficking.

In the *upper subsoil* (0.28–0.33 m depth) significant deformations can be proven in all six treatments. The level of the deformations varies. The high-loaded tire caused the highest compaction (2004 and 2005).

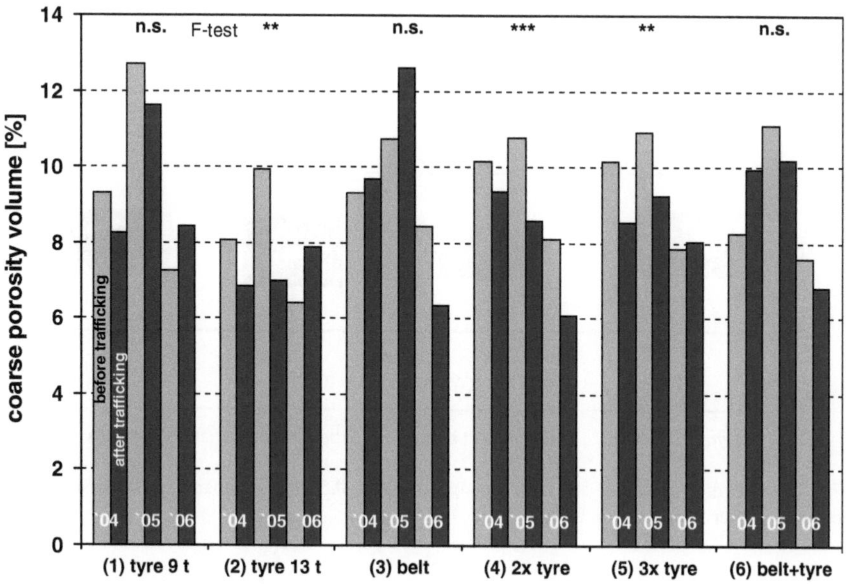

Fig. 7.4 Effects of different level of soil stress in the second subsoil (0.38–0.43 m): coarse porosity before (*gray columns*) and after (*black columns*) trafficking in 2004, 2005 and 2006

In the *second subsoil* (0.38–0.43 m depth) the analysis and the F-test shows that only treatment 2 with the highest load on one tire and the double and triple passes with high-loaded tires of treatments 4 and 5 caused significant deformations. These three treatments with significant compactions in the subsoil are also characterized by the highest mean contact area pressure.

There are no differences in the effects of both multiple pass treatments. It was not possible to separate the effects of multiple passes and wheel load.

The deformations in the upper subsoil and the second subsoil did not reduce the coarse porosity below the critical level of 5% (Werner and Paul 1999).

After the third and last trafficking in 2006 at all treatments with high wheel or belt loads, additional samples were taken at a depth of 0.48–0.53 m. The physical soil properties were identical before and after trafficking to a high level (coarse porosity about 10%). Figure 7.5 represents the comparison between the two treatments with tire load of 13 t and the rubber belt with 14 t weight.

7.3.5 Long-term Effects

At both depths the data analysis cannot prove a trend in the change of the physical soil properties over the whole investigation period. The measurements are characterized by large variations, which are obviously not only caused by the mechanical stress of the trafficking. What is remarkable is the repeated increase of the

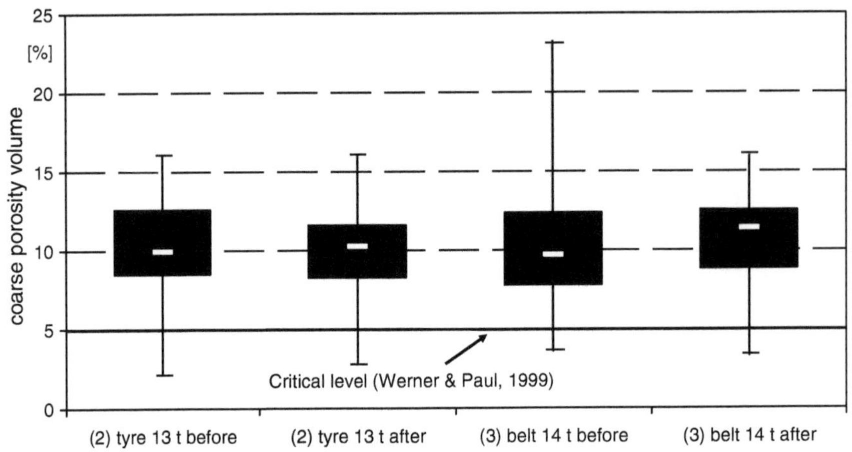

Fig. 7.5 Effects of high soil stress in the subsoil (0.48–0.53 m): comparison of coarse porosity before and after trafficking in 2006 only

coarse porosity in the periods between the sampling after trafficking and the sampling before trafficking a year later. Coarse porosity values in the second subsoil depth in 2006 should be carefully interpreted; trafficking and soil sampling took place under dry conditions. Therefore the sampling error might be higher than under wetter conditions. A part of the noise in the data can be explained by the fact that the area where the probes were taken was moved year by year along the defined tracks. In the area where the samples were taken in 2005 higher silt (+ 0.05 g g^{-1}) and lower clay (−0.05 g g^{-1}) contents in the subsoil were determined than in the neighboring areas of the 2004 and 2006 sampling.

Overall the time series is too short and the experimental design is not adapted enough to identify a trend proving an accumulation of deformation/compaction effects.

7.3.6 Effects on Water Infiltration and Yield

In addition the effects of trafficking on water infiltration and on yields (spring wheat and sugar beets in 2005, winter wheat and sugar beets in 2006 and spring oats in 2007) were determined, but less intensively than the abovementioned investigations. The significant effects of the different treatments on physical soil properties could not be found with water infiltration capacity and yields. In 2005 the treatments with double and triple passes with tires and high wheel loads showed lower infiltration rates and the treatment with the triple pass showed slightly lower winter wheat yields (Fig. 7.6). The reason for these effects seems to be compaction in the topsoil (crumb).

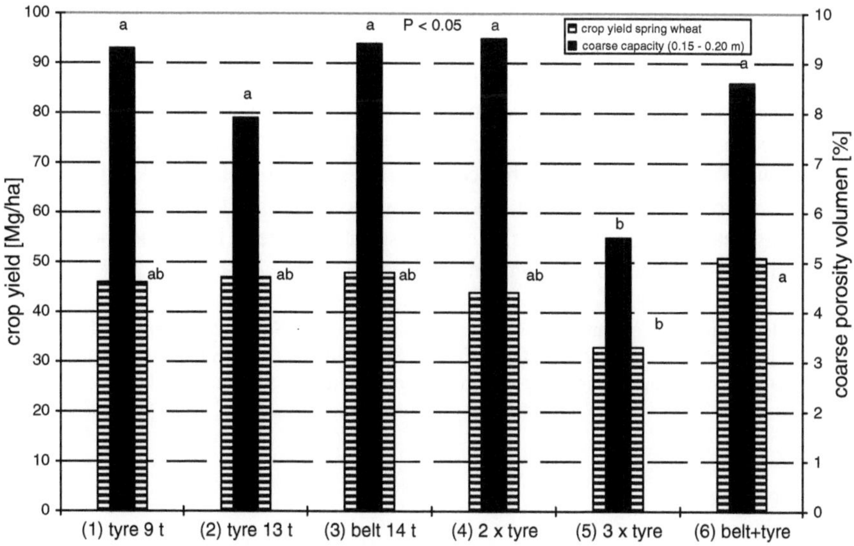

Fig. 7.6 Spring wheat yields in 2005 and bulk density in the topsoil (0.15–0.20 m)

7.4 Discussion and Conclusions

The modern rubber belt undercarriage investigated for large self-propelled harvesting machines caused no significant deformation in the subsoil, although the loads reached more than 14 metric tons. In combination with the results of the soil pressure measurements the risk of subsoil compaction can be assessed as lower compared to a wheel with a similar load (see also Ansorge and Godwin 2007). For high loads, rubber belt undercarriages are a soil-protecting alternative. But the documented (higher) topsoil compaction has to be taken into consideration.

The single pass of a tire with a wheel load of 13 metric tons (treatment 2) and both multiple passes with high loaded tires (treatments 4 and 5) caused significant deformations in the subsoil. The reduced deformation effects of the single tire pass with a reduced wheel load of 9 metric tons showed the soil-protecting potential of large-size high-volume radial tires not exceeding specific loads (in the case of the 1050/50R32 investigated, of 10 metric tons). The fact that all treatments caused significant deformations in the upper subsoil (former plow pan) proves that risk of soil compaction exists if the wheel loads exceed 9 metric tons using rubber belt undercarriages or large tires (see also Schäfer-Landefeld et al. 2004). The results also confirm that high wheel loads can cause soil deformation under the frequently cultivated topsoil at soil moisture levels that are typical and not extreme for agricultural (harvesting) applications. However, under the conditions of the investigation the depth of this impact was limited.

A concept to evaluate agricultural undercarriages and the soil stress they cause must also integrate the deformation of the topsoil. The results of the investigation indicate that multiple passes with high wheel loads can negatively influence the growth of the following crop if the moisture level in the topsoil exceeds a specific level.

An accumulation of repeated small deformation effects cannot be significantly proven after three years with three consecutive stress impulses under field conditions, but also cannot be rejected. To prove or disprove long-term effects on the soil caused by heavy agricultural machines, investigations over a longer period and with an specifically adapted design would be needed.

Acknowledgements We want to thank the German Union of Sugar Producers for funding the research projects in the years 2004–2007. We also want to thank the companies Grimme, Holmer, and Ropa for supporting the field experiments by making their equipment available.

References

Ansorge D, Godwin R (2007) The effect of tires and a rubber track at high axle loads on soil compaction, Part 1: Single axle studies. Biosyst Eng 98:115–126

Bolling I (1986) How to predict soil compaction from agricultural tires. J Terramechanics 22:205–223

Chamen WCT, Alakukku L, Pires S, Sommer C, Tijink F, Weisskopf P (2003) Prevention strategies for field traffic-induced subsoil compaction: a review. Part 2. Equipment and field practices. Soil Till Res 73:161–174

Geischeder R, Brandhuber R, Demmel M (2008) Potential of low ground pressure undercarriages to prevent soil compaction. AgEng Paper No. 1177044

Gysi M (2001) Compaction of a Eutric Cambisol under heavy wheel traffic in Switzerland: Field data and a critical state soil mechanics model approach. Soil Till Res 61:133–142

Håkansson I (2005) Compaction of arable soils. Incidence – consequences – counter-measures. Reports from the devision of soil management 109, 2005, Swedish University of Agricultural Sciences, Uppsala

Keller T, Arvidsson J (2004) Technical solutions to reduce the risk of subsoil compaction: effects of dual wheels, tandem wheels and tire inflation pressure on stress propagation in soil. Soil Till Res 79:191–205

Schaefer-Landefeld L, Brandhuber R, Fenner S, Koch H-J, Stockfisch M (2004) Effects of agricultural machinery with high axle load on soil properties of normally managed fields. Soil Till Res 75:75–8

Thangavadivelu S, Arnes P, Slocombe J, Higgins J, Stone L (1992) Soil response to track and wheel tractor traffic. Paper No. 921585, American Society of Agricultural Engineers, St. Joseph, Michigan, USA

Tijink FGJ, van der Linden JP (2000) Engineering approaches to prevent subsoil compaction in cropping systems with sugar beet. Adv GeoEcology 32:442–452

Werner D, Paul R (1999) Kennzeichnung der Verdichtungsgefährdung landwirtschaftlich genutzter Böden. In: Wasser und Boden. vol 51, Paper 12, pp 10–14

Chapter 8
Controlled Traffic Farming

G.D. Vermeulen, J.N. Tullberg, and W.C.T. Chamen

8.1 Introduction

Efficient mechanisation is a major factor underlying high productivity and low cost of agricultural production. Efficiency has generally been associated with greater work rates, achieved by using equipment of greater power and weight. However, where agriculture has become highly mechanised, the soil is regularly compacted by machinery wheels. To alleviate this traffic-induced soil compaction, the soil is usually loosened before growing the next crop. This cycle of compaction and tillage is a major inefficiency of current mechanised agriculture, particularly when the soil is tilled to greater depth. Not only does the present system waste fossil energy in soil compaction and re-loosening (Tullberg 2000), it also results in soil structure deterioration (Boels et al. 1982; Soane and van Ouwerkerk 1994). Considerable research has been undertaken to investigate the extent of traffic-induced soil damage under different tillage regimes.

A number of solutions have been proposed to avoid this cycle of soil compaction and tillage. Most often reduction of tillage, rather than traffic, has been seen as the solution. Although reduced tillage has provided major benefits in many parts of the world, the primary cause of compaction problems remains wheel traffic. This was identified as long ago as 1966 by Arndt and Rose (1966) who noted that "excessive traffic necessitates excessive tillage". A number of mechanisation options have

G.D. Vermeulen (✉)
Plant Research International, P.O. Box 616, 6700, AP Wageningen, the Netherlands
e-mail: bert.vermeulen@wur.nl

J.N. Tullberg
University of Queensland and CTF Solutions, 8 Hakea Crescent, Chapel Hill, Queensland 4069, Australia

W.C.T. Chamen
CTF Europe Ltd, Bedfordshire, UK

A.P. Dedousis and T. Bartzanas (eds.), *Soil Engineering*, Soil Biology 20,
DOI 10.1007/978-3-642-03681-1_8, © Springer-Verlag Berlin Heidelberg 2010

been reported to reduce the incidence of compaction under wheels (Soane et al. 1981; Chamen et al. 1992; Larson et al. 1994; Chamen et al. 2003). These include lowering vehicle ground pressures (Söhne 1953; Soane et al. 1981; Rusanov 1991; Vermeulen and Klooster 1992; Erbach 1994; Vermeulen and Perdok 1994), restricting axle loads (Hakansson and Petelkau 1994) and controlled traffic (Taylor 1983; Chamen et al. 1992; Taylor 1994). This chapter focuses on controlled traffic farming as a means of avoiding soil compaction over most of the field area. The chapter's objectives are to explain the principle of controlled traffic and its practice in Australia, UK and the Netherlands and to provide an overview of CTF research results from around the world.

8.2 Controlled Traffic Farming Systems

The Controlled Traffic concept was initiated in the USA around 1950 to increase crop yields by reducing soil compaction (Taylor 1983). In controlled traffic farming (CTF), equipment is adapted so all field operations are supported from permanent traffic lanes to allow optimum production from wide, non-trafficked crop beds. In practice it means repeated use of the same wheel tracks for all operations using a precise machinery guidance system. This can be achieved with specifically designed wide-span machines (gantries), but most systems are presently based on modified conventional agricultural equipment.

The gantry provides the optimal controlled traffic solution (Taylor 1994). Gantries are machines whose implements normally work within the span of their widely spaced pairs of tandem wheels or tracks (Fig. 8.1). The track width (the distance between the left and right wheel centres) of gantries varies from 4 to 12 m

Fig. 8.1 Dowler gantry with a track width of 12 m (Source: Experimental Farm Oostwaardhoeve, Slootdorp, the Netherlands)

and some 21 m wide units have been used. Such vehicles minimise traffic lane area. Another important advantage of gantries over machinery with a narrower track width is increased stability, leading to excellent potential for high precision operations. Though the use of gantries is reported to have great potential (Chamen et al. 1994a, b), a shift to wide span is a dramatic change for farming and will not come about by normally accepted methods. The existing industry has to be convinced that it is both practical and economically viable before it will be persuaded to invest at the significant level required.

Various types of tractor-based CTF systems have been investigated and used in practice. These range from rigid but efficient systems, in which all implement widths are the same and machines have identical track widths and tread or belt widths, to more flexible systems with two or more track widths and various tread or belt widths. In all cases, implement widths are constrained to a specific base value or a multiple of this, defined by the CTF system being used. Maximum advantage can be achieved only in a full system with permanent traffic lanes and permanently non-trafficked beds in which CTF is combined with no tillage. Partial systems are those where some heavy wheels are not restricted to permanent traffic lanes. One example is the seasonal controlled traffic farming (SCTF) system where traffic lanes are not used for primary tillage or harvesting, but for all operations in between.

CTF adoption has been facilitated by the development of 0.02-m precision machine guidance systems using RTK-DGPS (Real Time Kinematic Differential Global Positioning Systems) (Dijksterhuis et al. 1998). Accurate guidance and easy traffic lane installation and maintenance is possible with these systems, which often provide a topographical mapping facility to assist with layout.

8.2.1 Australian Systems

The 3-m track width system commonly used in Australian dryland grain production has been dictated by compromise between the limitations of harvesters, usually the most difficult machines to change, and tractors. Current large grain harvesters cannot operate at a track width significantly less than 3 m, while tractors cannot easily be modified to a track width much greater than 3 m. The rear of the "long axle" models of US-made tractors adjusts easily to 3 m, and small companies have produced front axle modifications. These were often simple wheel-rim extensions, but at least one company offers a highly professional service in extending driven front axles to 3 m (Fig. 8.2). The warranty situation of tractors modified to 3 m was often unclear, but warranted 3-m models were introduced by a major manufacturer in 2003, and other manufacturers have followed suit.

A typical example of such a 3-m track width system is illustrated in Fig. 8.3. This 3 m/9 m/27 m (track width/seeding and harvesting width/spraying width) system is probably the most common CTF system in use in Australia now, but 3 m/12 m/36 m systems are not uncommon on larger properties. Crops are

Fig. 8.2 Commercially available extensions of the front axle of a standard tractor to a track width of 3 m in Australia (Source S. Dick, Tasweld P/Ltd, Toowoomba)

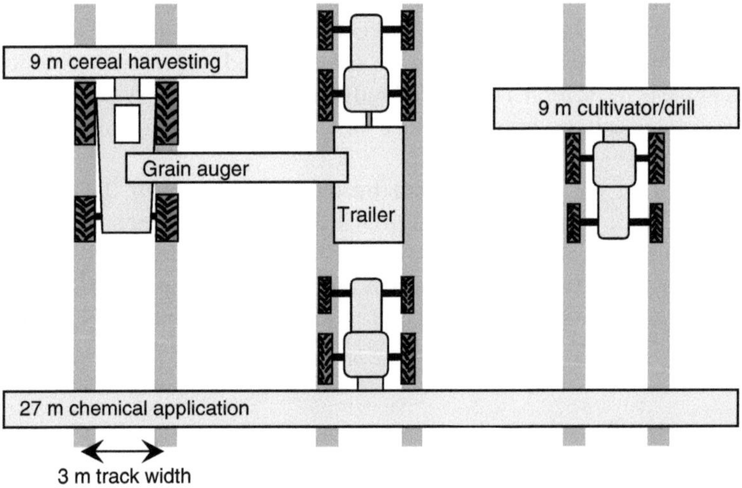

Fig. 8.3 3 m/9 m/27 m CTF system with an implement base width of 9 m and a three times multiple to give 27 m for chemical applications. The track width is 3 m

produced without tillage, which is much easier in CTF, and traffic lanes are normally not seeded. Using 0.5-m width tyres or less, only 8–12% of the field area is tracked in these systems.

Controlled traffic is now generally accepted as good practice within the Australian grains industry, although the importance of controlling all traffic is not always appreciated so estimates of 15% of grain production under CTF probably includes systems where harvest traffic is still random. Other systems have been based on track width values of around 2 m in irrigated crops, particularly cotton, and "raised bed" dryland systems, where grain harvester track width is sometimes extended to span two beds. Despite some excellent examples of 3-m production, the sugarcane industry is still dominated by 1.5 m cane rows and single-row harvesting with harvesters of 1.85 m track width, always accompanied by haulage units of variable track width and inexact steering. Unsurprisingly, soil degradation is a major problem for this industry where flexibility is limited by a perennial crop and harvesting is dominated by some hundreds of contractors. Horticultural production is faced with similar problems, and industry-wide co-operation will be needed to achieve solutions.

8.2.2 Flexible CTF Systems for Combinable Crops in UK

The most efficient controlled traffic system is one based on a common track width for all the machines, but unfortunately practicalities do not always allow this. In the UK and probably much of Europe, matching all machines to the wide track width of a combine harvester (which is only used for a very limited period during the year) is often impractical because of the narrow farm tracks and roads on which other vehicles in the system must run for much of the year. To overcome this problem, flexible systems incorporating different track widths and implement widths have been developed. Examples are the OutTrac (Fig. 8.4) and HalfTrac (Fig. 8.5) systems (CTF Europe 2008). OutTrac is very similar to the Australian system, other than using two different track widths, which is also the basis of the HalfTrac system. The implement width in the HalfTrac system is equal to the smallest track width (base implement width) or a multiple of this. Although these flexible CTF systems generally increase the tracked area, it is still considerably less than

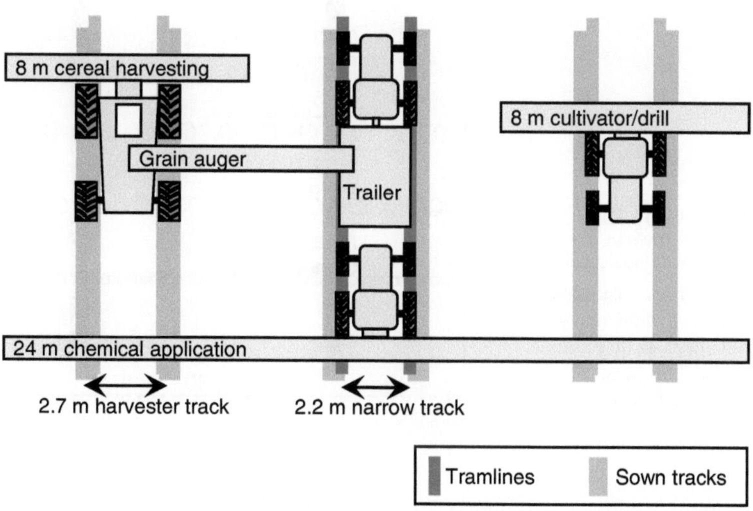

Fig. 8.4 An OutTrac controlled traffic farming system that uses two track widths, 2.2 m and 2.7 m, and a base implement width of 8 m. The tracked area in this example is 22.5 % of the field (harvester tyre-tread width of 0.80 m and tractor tread width of 0.50 m)

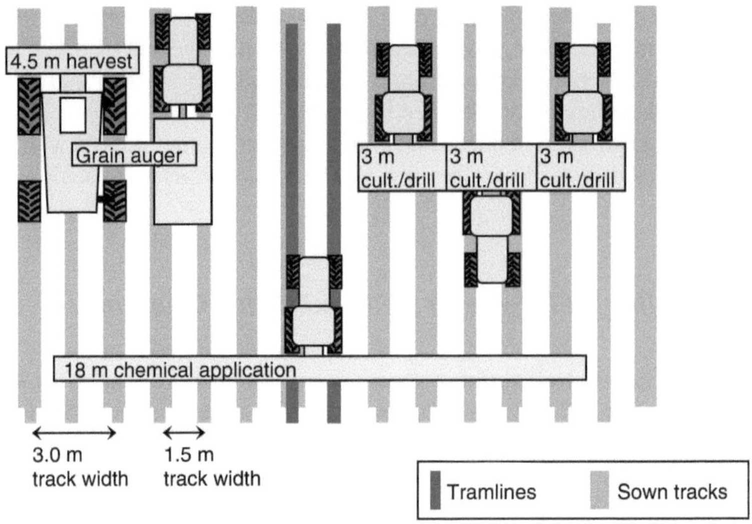

Fig. 8.5 A HalfTrac controlled traffic farming system with two standard track widths, one exactly half of the other. Implement width is constrained to the smallest track width or a multiple of this. The illustration shows a 3 m cultivator/drill, a 4.5 m harvester and an 18 m chemical applicator that together track 39% of the area (harvester tyre-tread width of 0.65 m and tractor tread width of 0.45 m)

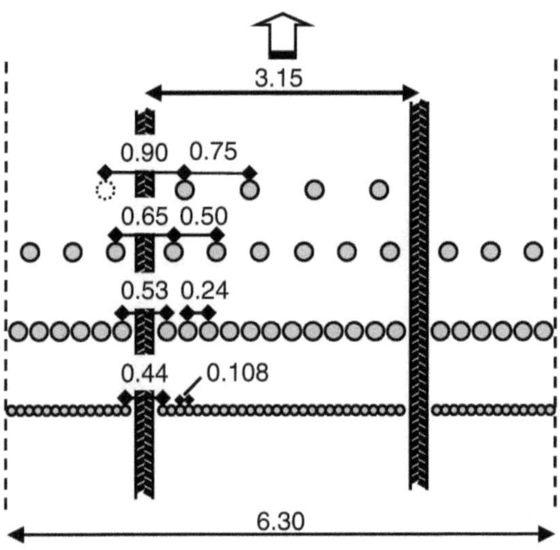

Fig. 8.6 Lay-out of traffic lanes and crop rows in a SCTF system suitable for various row crops (distances in m). The track width and the base implement width in this system is 3.15 m. Most implements are 6.30 m wide

conventional practice, particularly when considered over a number of years. Several different CTF systems are being tried on farms in the UK. No specific system is commonly used, but unlike Australian CTF systems for grains, traffic lanes are normally cropped when not used for chemical applications.

8.2.3 Dutch SCTF System for Organic Vegetable Farming

In the Netherlands, CTF has been adopted on about ten organic vegetable farms, growing a variety of crops. On most farms, an implement base and machine track width of 3.15 m with a fixed traffic-lane width of 0.30 m was chosen (Fig. 8.6). The conventional 0.50 and 0.75 m row spacings were maintained so that most of the conventional implements needed little adaptation. The traffic lanes are not cropped and the distance between the rows next to the traffic lanes has been increased to avoid compaction effects on these rows and provide more space for the tyres or tracks. The smaller row distances for cut crops were adjusted slightly to maintain about the same number of plants per hectare as on conventionally managed fields.

The system is seasonal because it was not economically feasible to modify harvesters and trailers for a 3.15-m track width and 0.30-m traffic lane width. Beds can be harvested with existing 3-m working width machines, but their wheels are not on the traffic lanes. Annual ploughing is a standard weed control practice in organic systems, but the disruption of the traffic lanes determines the "seasonal" nature of this CTF system. Interestingly, much earlier CTF experiments in the Netherlands developed a special mouldboard plough to limit tillage to the planting bed in 3-m and 3.3-m track width systems (Perdok and Telle 1978; Lamers et al. 1986).

The main difference between SCTF and conventional farming is omission of pre-sowing and in-crop field traffic in plant beds. This reduces compaction of the crop bed, but the traffic lanes reduced the area available for the crop by about 5% (0.15 m out of 3.15 m base width) for the wider row spacings, and by about 10% (0.30 m out of 3.15 m base width) for the narrower row spacings.

8.3 Impact of Controlled Traffic Farming

From the early work on CTF in the USA (Cooper et al. 1969; Dumas et al. 1975), it became clear that CTF increased crop yields as well as bringing other benefits. Taylor (1983) reported that the establishment of permanent crop zones and traffic lanes had reduced the need for deep tillage, tillage draught, the number of field operations and the amount of power required per hectare. It improved tractive efficiency and flotation, improving timeliness of operations and crop quality, while water infiltration rates and storage capacity increased and hence runoff decreased. These characteristics make CTF highly compatible with zero and reduced tillage systems, and it was suggested that it can be difficult to completely eliminate tillage without CTF. It is clear that CTF should be regarded as a complete soil compaction management system and not just a means of increasing crop yields (Taylor 1994). This section provides an overview of the impact of controlled traffic on the various aspects of crop production that determine the sustainability of production.

8.3.1 Soil Structure

CTF or SCTF generally lead to a soil structure with reduced bulk density and increased porosity in the growing season, compared with random traffic. This is the case when the soil is tilled annually (Chamen et al. 1990; Vermeulen and Klooster 1992; Vermeulen and Mosquera 2008) and when no tillage is applied (Dickson and Campbell 1990; Douglas et al. 1992). When no tillage is adopted along with CTF, bulk densities and penetration resistance increase much less for CTF than for random traffic (Dickson and Campbell 1990). Before adopting CTF and no tillage, soil loosening is advisable when the soil is severely compacted. Radford et al. (2007) found that, without tillage, it took a minimum of three years for a vertisol in Queensland, Australia to recover from compaction. This was reflected in terms of penetration resistance and yield reductions and in soil shear strength in the top 100 mm of the profile. McHugh et al. (2009) estimated that it would take about nine years for natural amelioration to penetrate to 1 m depth in a similar soil, and showed that most of the improvements in soil structure and plant available water achieved over four years of controlled traffic were destroyed by one pass of a tractor wheel with a wheel load of 2 Mg in typical seeding conditions (trafficable surface, moist at seeding depth)

8.3.2 Soil Biota

A good soil structure is a pre-requisite for satisfactory crop growth and adequate functioning of soil organisms. Soil structure supplies plant roots and soil organisms with "habitable pore space" and controls many processes, e.g. the transport of water, oxygen and nutrients. In turn, soil organisms can contribute to an optimum soil structure through aggregate formation and the creation of biopores. Soil organisms also play an important role in plant development, e.g. by supplying nutrients and by controlling pests and plant pathogens (Brussaard and Van Faassen 1994).

The effects of compaction on biological processes are complex. However, compaction of soil generally reduces biotic activity particularly in the case of roots, earthworms and other fauna. In the case of microbial activity, the emphasis tends to be changed from aerobic to anaerobic activity with compaction. The process of soil tillage is often disruptive and harmful to the ecosystems that form in the soil, suggesting a need for re-evaluation of tillage objectives and outcomes, given the deleterious consequences (Whalley et al. 1995).

As CTF improves soil structure and gives better opportunities for reduced tillage, one may expect positive effects of CTF on biotic activity. Pangnakorn (2002) found that earthworm numbers in random traffic no tillage and controlled traffic no tillage plots in dryland grain farming were three and eight times that of annually wheeled, tilled plots, respectively. Other soil macrofauna followed a generally similar pattern, but trends were not as obvious in microfauna.

In an experiment with annual ploughing for spring wheat and winter barley in Germany, Söchtig and Larink (1992) investigated bulk density, earthworm numbers and earthworm biomass in plots with zero traffic and plots with wheelings during seedbed preparation, harvesting, and stubble mulch tillage. They concluded that bulk density decreased from 1.49 to 1.32 Mg m^{-3}, the number of earthworms increased by 26% and the earthworm biomass increased by 12% for zero traffic, compared with the wheeling treatment.

Radford et al. (2001) reported that, compared with reduced tillage regimes with traffic and tillage under dry soil conditions, the numbers of macrofauna and earthworms were reduced by a factor of at least five in either:

- Annually wet-compacted soil (wheel load 10 Mg) without tillage; or
- Annually wet-compacted soil (wheel load 6 Mg) with frequent tillage.

They suggested that tillage and traffic under wet soil conditions is particularly harmful to earthworms for various reasons, e.g. because of direct mechanical damage as earthworms migrate to the topsoil during wet conditions and because high soil strength inhibits animal movement after wet-compaction.

8.3.3 Soil Water

8.3.3.1 Hydraulic Conductivity, Infiltration and Runoff

Changes in the soil structure by wheeling, tillage and settling after tillage also affect the movement of water in the soil. As the largest pores collapse first in the compaction process, saturated conductivity and infiltration rate decrease sharply with increasing compaction (Ankeny et al. 1990, Servadio et al. 2001). Meek et al. (1992) reported that natural channels in the soil (biopores) are particularly important to maintain high saturated conductivity and infiltration rate and they pointed out that tillage destroys these biopores. Reduced macroporosity and infiltration rate may lead to increased runoff and soil erosion in undulating areas and to ponding and decreased soil aeration on flat terrain.

Li et al. (2001) assessed the impact of compaction from wheel traffic on a clay soil that had been in CTF for five years. A tractor of 40 kN rear axle weight was used to apply traffic at varying wheel slip, with varying residue cover to simulate effects of traffic typical of grain production operations in the northern Australian grain belt. Simulated rainfall was used to determine infiltration characteristics. Wheel traffic significantly reduced time to ponding, steady infiltration rate, and total infiltration compared with non-wheeled soil, with or without residue cover. Non-wheeled soil had 4–5 times greater steady infiltration rate than wheeled soil, irrespective of residue cover. Wheel slip greater than 10% further reduced steady infiltration rate and total infiltration compared with that measured for self-propulsion wheeling (3% wheel slip) under residue-protected conditions. Where there was no compaction from wheel traffic, residue cover had a marked effect on infiltration capacity, with steady infiltration rate increasing proportionally with residue cover. Residue cover, however, had much less effect on infiltration when wheeling was imposed.

Tullberg et al. (2001) investigated traffic and tillage effects on runoff and crop performance on a heavy clay vertisol in Australia over a period of four years. Tillage treatments and the cropping programme were representative of broadacre grain production practice in northern Australia. CTF or annual whole-plot wheel traffic treatments (with a 100-kW tractor) were imposed on plots managed with no, minimum, and stubble mulch tillage. Rainfall/runoff hydrographs demonstrated that wheeling produced a large and consistent increase in runoff, whereas tillage produced a smaller increase. Treatment effects were greater when rainfall occurred on dry soil, but they were also large under intense rainfall events on wet soil. Mean annual runoff from wheeled plots was 63 mm (44%) greater than that from controlled traffic plots, whereas runoff from stubble mulch tillage plots was 38 mm (24%) greater than that from no tillage plots. Traffic and tillage effects appeared to be cumulative. The increased infiltration for CTF and no tillage was reflected in an increased yield of 16% compared with wheeled stubble mulch.

8.3.3.2 Soil Erosion

Rohde and Yule (2003) observed that runoff and sediment concentration can be reduced by surface soil protection through maintaining high levels of cover (stubble retention, minimum tillage, opportunity crops, cover crops), through high soil water deficits and by controlling traffic. These effects were clear from central Queensland data where plots double-cropped in controlled traffic no tillage with 41% residue cover produced 2.8 t ha^{-1} soil loss over 2½ years, but soil loss from otherwise similar single-cropped plots was 60% greater, and soil loss from annually wheeled, minimum tillage plots with 23% cover was 260% greater.

In addition to the reduction of runoff quantity and sediment concentration, measured on small plots, runoff management is necessary for large-scale, on-farm control of water erosion (Yule 1998). Runoff management has traditionally been achieved by using contour banks to reduce slope length, but this does not prevent rilling between contour banks. This is increased in severity when management operations are carried out "on the contour", because planting furrows, crop rows and wheel tracks all tend to concentrate runoff at the lowest point, increasing the area contributing runoff to each rill. In a field-scale experiment, Rohde and Yule (2003) demonstrated that downslope layouts (550 m long at 1% slope) developed no rills and were stable under rainfall events of up to 110 mm, with 15 min intensity levels of up to 66 mm h^{-1}. Erosion rates were low and responded to cover and antecedent water content. The data suggested that cover levels of 50% dramatically reduce runoff and erosion, but that suspended sediment concentration was reduced by only about 30%. Greater cover should further reduce suspended sediment load, which is important because this moves long distances into rivers, carries enhanced levels of nutrients and pesticides, and has a large environmental impact. Advocacy of "downslope", rather than the traditional "contour" operation patterns has been controversial, but proved effective for minimising soil loss (Yule 2008).

Field plot measurements of runoff and soil loss, together with replicated rainfall simulation experiments, were used by Wang et al. (2008) to assess effects of tillage, traffic and residue cover under maize monoculture on sloping land on the loess plateau of northern China. Runoff was reduced by avoiding wheel-induced soil compaction, maintaining maximum residue cover and minimising tillage. Soil loss appeared to be directly related to runoff. The positive effects of avoiding compaction, even by relatively light equipment, were greater than the effects of 70% residue cover, which in turn were greater than those of avoiding tillage. Compaction effects of small-scale farm equipment on loess in China appeared to be of the same order of magnitude as those of large-scale farm equipment on vertisols in Australia, at least in terms of their impact on runoff.

8.3.3.3 Drainage

In humid climates, compaction and the associated low water conductivities may lead to prolonged periods with ponded or wet soil. During these wet periods, oxygen diffusion is low and shortages of oxygen may occur in the soil (Boone 1986). Oxygen is needed for the proper functioning of plant roots and other soil biota. In the Netherlands, as a rule of thumb, 10% macroporosity (porosity at −10 kPa soil water matrix potential) is needed on clay soil for undisturbed root growth (Bakker and Hidding 1970) and 15–20% on sandy soils (Boone et al. 1986). Because CTF increases soil porosity, one may assume that the oxygen supply in the soil is less of a limiting factor for crop growth under wet soil conditions. To develop the full potential of CTF in terms of drainage, Lamers et al. (1986) suggested subsoiling to increase cropping zone porosity before adoption of CTF. Yule (1998) subsequently noted that "downslope" CTF layouts provide excellent drainage, so traffic lanes become trafficable more rapidly after rainfall events.

8.3.3.4 Workability

Impeded drainage, associated with soil compaction and prolonged periods with ponded or wet soil, may also restrict the number of workable days (Van Wijk and Feddes 1986). Therefore, CTF may be expected to have more workable days than conventional random traffic farming. Another advantage of CTF is that the wheels of machinery only drive on the traffic lanes and not on the cropping zones.

Spoor (1997) in a study of sugar beet establishment found controlled traffic increased the days available for drilling by up to 14 depending on soil and season. This was helped in part by more accessible traffic lanes that reduced towing forces by up to 30% depending on soil type and condition. Increased timeliness makes early planting possible, which often results in yield increases. Vermeulen and Klooster (1992) calculated the number of suitable work days for tillage and traffic in high ground pressure, low ground pressure and zero traffic and found a significant increase in the number of suitable work days for zero traffic.

Increased timeliness is particularly useful when weed control has to be performed by mechanical means, such as in organic farming. More workable days, and thus improved timeliness, means that the weeds can be controlled at a suitable moment. Another advantage of CTF in relation to mechanical weed control is that CTF provides a uniform flat cropping zone, free from wheel ruts, but with sufficient loose soil to allow very efficient operation of sweep cultivators.

In Australia, improved timeliness provides greater cropping opportunities, including double cropping where it was not possible before. This has allowed additional economic crops in some areas, and the inclusion of cover crops for improved soil organic matter levels and soil structure in others (Bowman 2008).

8.3.4 Crop Yield

The factors associated with compaction that reduce the growth potential of crops include water infiltration, plant available water capacity, oxygen supply and dentrification. Poor rooting of the crop due to excessive soil strength may exacerbate these influences. To what degree these factors will become limiting depends on interactions between crop type, soil type, weather conditions and the degree of compactness of the soil (Lipiec and Simota 1994; Lindstrom and Voorhees 1994). Seed-zone firming is generally accepted as a means to improve seed–soil contact and is not regarded as soil compaction.

In most experiments where yields from compacted soil are compared with the yield of loose soil, yields are negatively affected by compaction. These negative outcomes are not confined to specific crops, soils, climates or farming systems. In a review, Chamen (2005) reported that negative responses to compaction for 15 different crops ranged from 2–81% with wheel loads from 1 to 10 Mg. There were only three instances in a total of 79 when a positive response to compaction was recorded.

When conventional systems and CTF systems are compared in practice, the soil in conventional systems may not be 100% wheeled before seeding and low ground pressure may be used. On the other hand, the soil in CTF systems may be somewhat compressed by depth wheels or other elements of the implements. CTF may also use extra non-cropped space for the traffic lanes, which reduces the yield per ha. The differences in yield may therefore be smaller than in dedicated experiments. Nevertheless, full-scale experiments show that yield responses are generally positive.

Yields in extensive grain production in Australia are normally limited by moisture availability. Drought is common, but can be punctuated by high-intensity rainfall events, so factors such as the improvement in infiltration rates and plant available water capacity provided by CTF are important, together with compatibility with no tillage and maintenance of maximum standing crop residue. Careful layout of traffic lanes to provide rapid drainage and efficient logistics is essential for good crop performance. Under these conditions, traffic-related yield differences in grain were significant in all crops (Tullberg et al. 2001). Mean yield of controlled traffic plots was 523 kg ha^{-1} (14%) greater than that of wheeled plots. Except for the first year in winter wheat, the mean yield of no-till plots was 2–8% greater than that of stubble mulch plots for all crops, but these differences were rarely significant. It is important to note that the yield data came from traditional, side-by-side plots in which all treatments were planted and harvested at the same time. Improved timeliness is probably the major factor in the much greater "system" response, including crop yields, achieved when controlled traffic is applied to large-scale farming practice (Bowman 2008).

Yields in the temperate climate zone of Europe are normally not limited by moisture availability. However, limited supply of oxygen to roots and soil fauna may be a problem in compacted soil, particularly during wet periods. Therefore, good drainage is essential for good yields.

In a fully implemented CTF system with subsoil loosening before adoption of CTF, an average yield increase of about 10% was achieved for crops with a relatively weak root system (seed potato) on a sandy clay loam soil, relatively vulnerable to compaction. No yield increase was found for sugar beet on a young polder soil with a loam texture (Lamers et al. 1986). Vermeulen and Klooster (1992) reported that crop yields of root crops (sugar beet, onion and potato) on sandy clay loam soil increased by 9% compared with conventional traffic. Similarly, reported yield increases were 7% for sugar beet in Germany (Sommer and Zach 1992) and 18% for potato in Scotland (Dickson et al. 1992). Data on yields of winter grains in England, the Netherlands and Germany showed a yield response in the range of −9% to +25%, indicating that loose soil conditions can reduce yield on occasions (due to reduced access to manganese), while the yield of winter barley in Scotland showed positive response compared with a system with conventional, high-ground-pressure equipment (Chamen et al. 1992, Chamen and Longstaff 1995).

In a SCTF system applied in organic vegetable farming, yields increased significantly compared with random low-ground-pressure traffic in green pea (31%), spinach (15%) and planted onion sets (10%), but not in carrot and in direct-sown onion (Vermeulen and Mosquera 2008). The yield of ryegrass from permanent grassland increased by 16% when CTF was used rather than conventional high-ground-pressure equipment and random traffic in Scotland (Douglas et al. 1992).

8.3.5 Energy Use and Emission of Greenhouse Gases

Removing vehicle-induced compaction from the cropped area reduces tillage energy requirements dramatically as well as the need for tillage per se. Energy savings of up to 70% have been recorded within particular cropping systems (Chamen et al. 1992; McPhee et al. 1995). Energy savings include the savings from fewer operations, shallower depths of operation and lower pulling force requirements of the implements involved. Within a controlled traffic regime, there are also around 13% savings from the improved tractive efficiency of running on compacted traffic lanes (Lamers et al. 1986). On non-trafficked soils a pulling force reduction of 37% has been recorded for tillage at around 100 mm depth (Chamen and Longstaff 1995), 20–30% for tillage at 200 mm depth (Canarache et al. 1984; Dickson and Campbell 1990) and 18% for loosening at 0.55 m depth (Chamen and Cavalli 1994). Overall, random trafficking can increase the power used for a given operation by around 100% (Tullberg 2000).

In a review, Mosquera et al. (2005) reported that soil compaction has been observed to increase N_2O emissions by 20–50% on average. The effect of soil compaction on N_2O emissions is generally higher in clay soils, and lower in sandy soils. Soil compaction has also been reported to reduce the ability of soils to absorb atmospheric CH_4 by 60% on average (range: 30–90%). The effect of soil compaction on CH_4 fluxes is such that, in some cases, net CH_4 sinks are transformed into net emission sources.

Vermeulen and Mosquera (2008) demonstrated that these effects occurred in practice when the Dutch SCTF system with annual ploughing was compared on-farm with conventional random traffic farming (RTF). In the experiment, the average air-filled porosity increased from 16 to 19% (depth 0.02–0.07 m) and from 10 to 13% (depth 0.10–0.15 m) respectively for RTF and SCTF. SCTF resulted in a significant reduction of N_2O emissions (by 20–50% compared to RTF). For CH_4, application of the SCTF system resulted in increased CH_4 uptake (by a factor of 5–20) in three fields and in lower (but not significant) CH_4 emissions (by a factor of 4) on the fourth field, compared with RTF.

In attempting to quantify the overall greenhouse gas impact of common Australian cropping systems, Tullberg (2009) made the distinction between input-related emission sources which are easy to account for in a known system (i.e. fuel, machinery, herbicides and fertiliser), and soil emissions, where our knowledge is much less certain. From his analysis he concluded that input-related emissions from typical examples of tillage-based stubble mulch, random traffic no tillage and controlled traffic no tillage cropping systems occurred in the ratio of 100:97:78, but the ratio for soil emissions was 100:134:66. The magnitude of soil emissions was substantially greater than that of input-related emissions, but these values were seen as highly speculative, being based on research results obtained under very different circumstances (e.g. Vermeulen and Mosquera 2008).

A recent review of available data by Rochette (2008) has confirmed the importance of drainage and aeration to nitrous oxide emissions, indicating that no tillage, compared with tillage, resulted in mean N_2O emissions that were 0.06 kg N ha^{-1} lower, 0.12 kg N ha^{-1} higher and 2.00 kg N ha^{-1} higher in soils with good, medium and poor aeration, respectively. These findings agree with the data of Ball et al. (2008), who stated for ploughed and no-tilled soil that the production and emission of N_2O were strongly influenced by the soil physical environment, the magnitude of the water-filled pore space and continuity of the air-filled pore space in particular. Controlled traffic generally results in a major improvement in soil porosity,

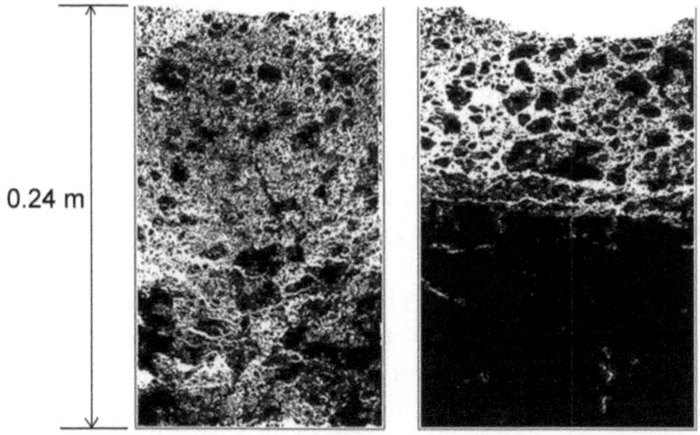

Fig. 8.7 Binary images of soil structure illustrating the 0–0.24 m depth pore system in non-wheeled CTF (*left*) and wheeled (*right*) soil under a zero tillage regime on a vertisol in Australia. Treatments were applied annually for four years after cereal harvest using a tractor with 4 Mg axle-load and 100 kPa tyre pressure. (Images supplied by D. McGarry, Queensland Department of Natural Resources and Water)

particularly under no tillage (Fig. 8.7). The effect of adopting no tillage along with CTF resulted in a limited increase in bulk density in Scotland (Dickson and Campbell 1990).

Pangnakorn (2002) measured mean bulk density (0–0.15 m) at 6-month intervals over two years in cropped vertisol plots that had been consistently tilled or non-tilled, wheeled or non-wheeled for six years. No difference was detected between tilled and non-tilled plots, regardless of wheeling treatment, but the mean bulk density of wheeled plots was 15% greater than that of non-wheeled plots. McHugh et al. (2009) also noted a bulk density difference >10% in the top 10 cm of nearby plots after two years of zero tillage controlled traffic. Greater infiltration rates in controlled traffic no tillage plots suggests that the reduced bulk density is associated with greater continuity of porosity, which might be expected to reduce nitrous oxide emissions.

8.3.6 Practical Aspects and Economics

The practicability and positive economic aspects of CTF for dryland grain production in Australia have been demonstrated by large-scale farmer adoption, and the area presently managed under some type of controlled traffic has been estimated at around 2 Mha (Tullberg et al. 2007). Detailed analysis of the economics of controlled traffic is rare, but after interviewing 16 farmers in one area and noting their estimates of conversion costs and productivity impacts, Bowman (2008) calculated that the impact on total gross margin was an improvement of 68%. This resulted from a combination of increased yield and reduced costs, with simultaneous benefits in environmental performance.

Australian growers have frequently commented on the reduction in "time spent farming" with the adoption of no-tillage CTF, and on the greater ease of management once a CTF system is fully operational. Essential practical issues include the importance of setting the grain harvester header as high as possible and the use of an effective straw spreader. Seeding problems in no tillage can be reduced with a facility to offset seeding equipment slightly from the previous season's row. In some circumstances, growers have been able to get a second crop by replanting almost directly behind the harvester. Other benefits achieved with 2-cm precision RTK-DGPS guidance systems in CTF include the use of narrow-band application of expensive agricultural chemicals, and precise in-crop placement of fertilisers.

Traffic regulations can be a factor limiting the application of CTF machinery. The total width of agricultural machinery is often restricted to a legislated maximum when the machinery has to be operated both in the field and on the road, as is the case in most EU countries. As mentioned earlier (Sect. 8.2.2), wide machines are also often impractical and therefore CTF systems with different track gauges and implement widths have been developed, as in the UK (Figs. 8.4 and 8.5). This allows the use of standard equipment, even if it has to be very particularly matched with other equipment in the system. Most farmers are practically minded and once they have grasped the principles of CTF, can design their own, often unique solutions that may require only minor adaptations to existing equipment.

One of the greater challenges however is the length of the unloading auger on combine harvesters, which must reach close to the centre of the adjacent permanent wheel tracks. Some growers have responded to this by building sub-hoppers on the side of grain trailers that elevate and transfer material to the centre of the container. Dealing with baled residues also creates a challenge, but as with all machinery systems, rising demand often results in innovative and cost-effective solutions.

The economics of CTF on UK farms with crops that can be combine-harvested were reviewed by Chamen et al. (1994b). In their study, they concluded that, in order to justify the use of the zero-traffic systems, yield increases and larger scale farms (400–500 ha) would be needed. More recent work, based on the introduction of high-accuracy machinery guidance systems, suggests that smaller farms could benefit equally. The cost of inputs has also risen dramatically and recent but unpublished studies on a hypothetical 400 ha farm suggest that a CTF system based on RTK guidance and 27% tracked area would have 7% lower operating costs compared with a non-inversion tillage system, the latter using a less expensive satellite-based correction signal. With net yield responses taken into account for the CTF system, a 14% greater operating profit was predicted. Similar comparisons with a plough-based system were 15% lower operating costs and a 21% increase in profitability.

Although several years of controlled traffic research in the Netherlands using modified conventional tractors has shown several benefits (Lamers et al. 1986; Vermeulen and Klooster 1992), CTF was not considered an economically attractive option for arable farming in the Netherlands (Janssens 1991). The interest in CTF in the Netherlands was renewed when precise machinery guidance became available, based on RTK-DGPS. The first practical application of CTF in the Netherlands was in organic farming, where optimal soil structure is considered essential to obtain reasonable yields, where high-value vegetable crops are usually part of the rotation and where protection of the environment is a priority. Based on the cost of machinery needed on 50 ha and 200 ha organic farms for both systems, it was estimated by Vermeulen et al. (2007) that compared with RTF, SCTF is more profitable when the average crop yield increases by more than 1.6% in the case of a 50 ha farm and by more than 2.2% in the case of a 200 ha farm. These assessments were done without accounting for benefits from increased timeliness and more workable days. As the results suggest an average yield increase of 6–10%, SCTF is considered an economically feasible option in organic vegetable farming.

8.4 Conclusions

There are many benefits associated with CTF and they all help sustain farming systems and improve profit. Primarily the benefits are delivered by avoiding the cyclic compaction and loosening of the soil, thereby improving soil structure. This in turn lowers costs and increases crop returns, as well as improving system efficiency. Controlled traffic systems restore to cropped soils some of the environmental functions associated with soils under natural vegetation.

Lower costs and increased returns are mainly brought about by:

- Lower cost for soil tillage
- Higher yields
- Improved field efficiency
- More workable days
- Precision application of inputs.

Improved environmental functioning is mainly brought about by:

- Better infiltration and drainage of the soil
- Less water run-off and soil erosion
- Less denitrification and emissions of N_2O and CH_4
- Less use of fossil energy
- Potential to retain more organic matter and soil-living organisms.

References

Ankeny MD, Kaspar TC, Horton R (1990) Characterization of tillage and traffic effects on unconfined infiltration measurements. Soil Sci Soc Am J 54:837–840

Arndt W, Rose CW (1966) Traffic compaction of soil and tillage requirements. J Agr Eng Res 11:170–187

Bakker JW, Hidding AP (1970) The influence of soil structure and air content on gas diffusion in soils. Neth J Agric Sci 18:37–48

Ball BC, Crichton I, Horgan GW (2008) Dynamics of upward and downward N_2O and CO_2 fluxes in ploughed or no-filled soil in relation to water-filled pore space, compaction and crop presence. Soil Till Res 101:20–30
Boels D, Davies DB, Johnston AE (1982) Soil Degradation. Balkema, Rotterdam
Boone FR (1986) Towards soil compaction limits for crop growth. Neth J Agric Sci 34:349–360
Boone FR, Van der Werf HMG, Kroesbergen B, Ten Hag BA, Boers A (1986) The effect of compaction of the arable layer in sandy soils on the growth of maize for silage. 1. Critical matric water potentials in relation to soil aeration and mechanical impedance. Neth J Agric Sci 34:155–171
Bowman K (2008) Economic and environmental analysis of converting to controlled traffic farming. 6th Australian Controlled Traffic Farming Association conference, Dubbo, pp 61–68
Brussaard L, van Faassen HG (1994) Effects of compaction on soil biota and soil biological processes. In: Soane BD, van Ouwerkerk C (eds) Soil Compaction in Crop Production. Elsevier, Amsterdam, pp 215–235
Canarache A, Colibas J, Colibas M, Horobeanu I, Patru V, Simota H, Trandafirescu T (1984) Effect of induced soil compaction by wheel traffic on soil physical properties and yield of maize in Romania. Soil Till Res 4:199–213
Chamen WCT (2005) Controlled traffic farming – its benefits and realisation. Revista de la Facultad de Agronomía, La Plata 106 (1) Suplemento, Facultad de Ciencias Agrarias y Forestales, UNLP, Argentina, pp 107–129
Chamen WCT, Cavalli R (1994) The effect of soil compaction on mole plough draught. Soil Till Res 32:303–311
Chamen WCT, Longstaff DJ (1995) Traffic and tillage effects on soil conditions and crop growth on a swelling clay soil. Soil Use Manage 11:168–176
Chamen WCT, Chittey ET, Leede PR, Goss MJ, Howse KR (1990) The effect of tyre/soil contact pressure and zero traffic on soil and crop responses when growing winter wheat. J Agr Eng Res 47:1–21
Chamen WCT, Vermeulen GD, Campbell DJ, Sommer C (1992) Reduction of traffic-induced soil compaction: a synthesis. Soil Till Res 24:303–318
Chamen WCT, Dowler D, Leede PR, Longstaff DJ (1994a) Design, operation and performance of a gantry system: experience in arable cropping. J Agr Eng Res 59:45–60
Chamen WCT, Audsley E, Holt JB (1994b) Economics of Gantry- and Tractor-based Zero-Traffic Systems. In: Soane BD, van Ouwerkerk C (eds) Soil Compaction in Crop Production. Elsevier, Amsterdam, pp 569–595
Chamen WCT, Alakukku L, Pires S, Sommer C, Tijink F, Weisskopf P (2003) Prevention strategies for field traffic-induced subsoil compaction: a review. Part 2. Equipment and field practices. Soil Till Res 73:161–174
Cooper AW, Trouse AC Jr, Dumas WT (1969) Controlled traffic in row crop production. In: Proceedings 7th International Congress of Agricultural Engineering, Baden-Baden, pp 1–6
CTF Europe (2008) Excel file CTFsystems2008. http://www.controlledtrafficfarming.com. Cited 3 Feb 2009
Dickson JW, Campbell DJ (1990) Soil and crop responses to zero- and conventional-traffic systems for winter barley in Scotland, 1982–1986. Soil Till Res 18:1–26
Dickson JW, Campbell DJ, Ritchie RM (1992) Zero and conventional traffic systems for potatoes in Scotland, 1987–1989. Soil Till Res 24:397–419
Dijksterhuis HL, van Willigenburg LG, van Zuydam RP (1998) Centimetre-precision guidance of moving implements in the open field: a simulation based on GPS measurements. Comp Elec Agr 20:185–197
Douglas JR, Campbell DJ, Crawford CE (1992) Soil and crop responses to conventional, reduced ground pressure and zero traffic systems for grass silage production. Soil Till Res 24:421–439
Dumas WT, Trouse AC, Smith LA Jr, Kummer FA Gill WR (1975) Traffic control as a means of increasing cotton yields by reducing soil compaction. Paper No 75–1050 (unpubl.) Am Soc Agr Eng St. Joseph, MI 49085

Ehrbach DC (1994) Benefits of tracked vehicles in crop production. In: Soane BD, van Ouwerkerk C (eds) Soil Compaction in Crop Production. Elsevier, Amsterdam, pp 501–520

Hakansson I, Petelkau H (1994) Benefits of Limited Axle Load. In: Soane BD, van Ouwerkerk C (eds) Soil Compaction in Crop Production. Elsevier, Amsterdam, pp 479–499

Janssens SRM (1991) Rendabiliteit van verminderde bodembelasting: bedrijfseconomische evaluatie van een lagedrukberijdingssysteem. Verslag 127, Proefstation voor de Akkerbouw en Groenteteelt in de Vollegrond, PAGV, Lelystad

Lamers JG, Perdok UD, Lumkes LH, Klooster JJ (1986) Controlled traffic farming systems in The Netherlands. Soil Till Res 8:65–76

Larson WE, Eynard A, Hadas A, Lipiec J (1994) Control and Avoidance of Soil Compaction in Practice. In: Soane BD, van Ouwerkerk C (eds) Soil Compaction in Crop Production. Elsevier, Amsterdam, pp 597–625

Li YX, Tullberg JN, Freebairn DM (2001) Traffic and residue cover effects on infiltration. Austr J Soil Res 39:239–247

Lindstrom MJ, Voorhees WB (1994) Responses of Temperate Crops in North America to Soil Compaction. In: Soane BD, van Ouwerkerk C (eds) Soil Compaction in Crop Production. Elsevier, Amsterdam, pp 265–286

Lipiec J, Simota C (1994) Role of soil and climate factors in influencing crop responses to soil compaction in central and eastern Europe. In: Soane BD, van Ouwerkerk C (eds) Soil Compaction in Crop Production. Elsevier, Amsterdam, pp 365–390

McHugh AD, Tullberg JN, Freebairn DM (2009) Controlled traffic farming restores soil structure. Soil Till Res 104:164–172

McPhee JE, Braunack MV, Garside AL, Reid DJ, Hilton DJ (1995) Controlled traffic for irrigated double cropping in a semi-arid tropical environment. Part 2. Tillage operations and energy use. J Agr Eng Res 60:183–189

Meek BD, Rechel ER, Carter LM, DeTar WR, Urie AL (1992) Infiltration rate of a sandy loam soil: effects of traffic, tillage and plant roots. Soil Sci Soc Am J 56:908–913

Mosquera J, Hol JMG, Hilhorst MA (2005) Precise soil management as a tool to reduce N_2O and CH_4 emissions from agricultural soils: literature review. In: Proceedings of the International Conference "Non-CO_2 Greenhouse gases (NCGG-4)", Millpress, Rotterdam, pp 5-12

Pangnakorn U (2002) Effect of tillage and traffic on soil organisms. Unpublished PhD Thesis, University of Queensland

Perdok UD, Telle MG (1978) Eerste ervaringen met rijbanenteelt en draadgeleiding. Landbouwmechanisatie 29:383–387

Radford BJ, Wilson-Rummenie AC, Simpson GB, Bell KL, Ferguson MA (2001) Compacted soil affects soil macrofauna populations in a semi-arid environment in central Queensland. Soil Biol Biochem 33:1869–1872

Radford BJ, Yule DF, McGarry D, Playford C (2007) Amelioration of soil compaction can take 5 years on a Vertisol under no till in the semi arid subtropics. Soil Till Res 97:249–255

Rochette P (2008) No-till only increases N_2O emissions in poorly-aerated soils. Soil Till Res 101:97–100

Rohde K, Yule D (2003) Soil compaction and Controlled Traffic Farming research in Central Queensland. In: ISTRO 16, Brisbane. Proceedings pp 1020–1027

Rusanov VA (1991) Effects of wheel and track traffic on the soil and on crop growth and yield. Soil Till Res 19:131–143

Servadio P, Marsili A, Pagliai M, Pellegrini S, Vignozzi N (2001) Effects on some clay soil qualities following the passage of rubber-tracked and wheeled tractors in central Italy. Soil Till Res 61:143–145

Soane BD, van Ouwerkerk C (1994) Soil Compaction in Crop Production. Elsevier, Amsterdam

Soane BD, Blackwell PS, Dickson JW, Painter DJ (1981) Compaction by agricultural vehicles: a review II. Compaction under tyres and other running gear. Soil Till Res 1:373–400

Söchtig W, Larink O (1992) Effect of soil compaction on activity and biomass of endogenic lumbricids in arable soils. Soil Biol Biochem 24(12):1595–1599

Söhne W (1953) Druckverteilung im Boden und Bodenverformung unter Schlepperreifen (Pressure distribution in the soil and soil deformation under tractor tyres). Grundl Landtech 5:49–63 (in German)

Sommer C, Zach M (1992) Managing traffic-induced soil compaction by conservation tillage. Soil Till Res 24:319–336

Spoor G (1997) Benefits from traffic control. Landwards 52:2–5

Taylor JH (1983) Benefits of permanent traffic lanes in a controlled traffic crop production system. Soil Till Res 3:385–395

Taylor JH (1994) Development and Benefits of Vehicle Gantries and Controlled-Traffic Systems. In: Soane BD, van Ouwerkerk C (eds) Soil Compaction in Crop Production. Elsevier, Amsterdam, pp 521–537

Tullberg JN (2000) Wheel traffic effects on tillage draught. J Agr Eng Res 75:375–382

Tullberg JN (2009) avoiding soil compaction in CA: controlled traffic systems for mechanised conservation agriculture and their effect on greenhouse gas balances. Proceedings 4th World Congress on Conservation Agriculture, New Delhi, February 2009. Lead papers, pp185–190

Tullberg JN, Ziebarth PJ, Yuxia Li (2001) Tillage and traffic effects on runoff. Australian J Soil Res 39:249–257

Tullberg JN, Yule DF, McGarry D (2007) Controlled traffic farming – From research to adoption in Australia. Soil Till Res 97:272–281

Van Wijk ALM, Feddes RA (1986) Simulating effects of soil type and drainage on arable crop yield. In: Van Wijk ALM, Wesseling J (eds) Agricultural Water Management. Balkema, Rotterdam, pp 97–112

Vermeulen GD, Klooster JJ (1992) The potential of a low ground pressure traffic system to reduce soil compaction on a clayey loam soil. Soil Till Res 24:337–358

Vermeulen GD, Perdok UD (1994) Benefits of Low Ground Pressure Tyre Equipment. In: Soane BD, van Ouwerkerk C (eds) Soil Compaction in Crop Production. Elsevier, Amsterdam, pp 447–478

Vermeulen GD, Mosquera J (2008) Soil, crop and emission responses to seasonal-controlled traffic in organic vegetable farming on loam soil. Soil Till Res 102:126–134

Vermeulen GD, Mosquera J, van der Wel C, van der Klooster A, Steenhuizen JW (2007) Potential of controlled traffic farming with automatic guidance on an organic farm in the Netherlands. In: Stafford JV (ed) Precision agriculture '07. Skiathos, Greece, pp 473–481

Wang X, Gao H, Tullberg JN, Li H, Kuhn N, McHugh AD, Li Y (2008) Traffic and tillage effects on runoff and soil loss on the Loess Plateau of northern China. Austr J Soil Res 46:667–675

Whalley WR, Dumitru E, Dexter AR (1995) Biological effects of soil compaction. Soil Till Res 35:53–68

Yule DF (1998) Controlled traffic farming – the future. In: Proceedings of the National Controlled Traffic Conference, University of Queensland, Gatton, pp 6–12

Yule DF (2008) CTF and GNSS – Know Where You Are Farming. GNSS Technical Support Newsletter, Issue 30, Oct 2008, pp 4–6. Department of Sustainability and Environment, Melbourne, Victoria

Chapter 9
Subsoil Compaction: Cause, Impact, Detection, and Prevention

Gunnar Hoefer and Karl Heinrich Hartge

In memoriam of Karl Heinrich Hartge

9.1 Background

The successful increase of yields which was brought about by use of fertilisers was backed up by extensive research in the field of plant nutrition. The success became overwhelming for more than a century to such an extent that the importance of the soil water regime for any plant growth lost was the focus for scientists as well as for practical farmers. This mainly happened because in many areas with highly sophisticated agriculture the soil water regime is no serious bottleneck to cereal production. Another important reason is that crops were adapted to the water regimes a long time ago, as this was the most important factor from the earliest times for extending agricultural cropping, while the field of soil structure for a long time was not considered. So, when first signs of damage of soil structure and its impact on plant yields were reported, no general theory of soil structure was available, and the large number of field experiments conducted in recent decades, gave no clear answer to the cause of the problem. However, agreement existed that weight-induced soil compaction was greatly increased by the use of modern tillage and harvesting equipment because of its steadily increasing weight. Therefore, first definitions of compaction were seen as a biologically oriented problem under the keyword of "healthy and sick soil life" (Sekera 1951). However, opinions differed widely – coming from either soil science or the agricultural machinery community – on tolerable load on soils in terms of vehicle total weight, weight per axle, and contact pressure between tyre and soil. Effects of different tyre air pressure were investigated as well as different tyre profile constructions (Alakukku et al. 2003). In general, agreement existed that pedologic processes should not be included in this problem as they would not act as fast as the effects observed in connection with the

G. Hoefer (✉) and K.H. Hartge (†)
Institute of Soil Science, Leibniz Universität Hannover, Herrenhäuser Str. 2, 30419 Hannover, Germany
e-mail: hoefer@ifbk.uni-hannover.de

A.P. Dedousis and T. Bartzanas (eds.), *Soil Engineering*, Soil Biology 20,
DOI 10.1007/978-3-642-03681-1_9, © Springer-Verlag Berlin Heidelberg 2010

recent acceleration of development of heavier agricultural machinery. Nevertheless, water-logging of the root zone was the most important cause for reduction in grain yield, even if it occurred under semiarid climatic conditions and only during short time intervals during the growing period. Finally, a lack of a general background became the reason for the differing research approaches as well as differing methods to ameliorate subsoil compaction. Deep ploughing and subsoil loosening with differently formed tools were used, wherein heaving up of the soil surface was seen as a sign of the degree of success of such an action. In addition, terms were created and temporarily used like "over-loosening, non-compacted soil, successful versus unsuccessful amelioration".

9.2 Cause

9.2.1 Conflicting Paradigm or A Philosophical Approach

When discussing the aims of technological development and their various effects on human life and possibilities, usually a final consent is reached that all permanent developments aimed towards economising at first the most valuable sector of resources. From earliest time until now this was human manpower, which might be used for providing raw material as well as working on it to produce the final result of the whole process. Losses at every separate step of the process were taken into account as long as they did not add up to outweigh the value of the expected use. As soon as one of the steps in the production line failed to contribute to this process, it would be exchanged or in severe cases the whole production line was abolished. Right from the early days of human technologic history the demand in manpower was this factor, so every step in the production was tailored to give optimal results until no further progress could be obtained. In the end, human manpower in the most general sense of the term was the final limit to further use or further experiments to increase its effectiveness. In the evolution of plant crop production, lines of technology were particularly long-lived. Their development was slow and their behaviour conservative compared to all other fields of handicraft. The reason is, as we know and frequently forget, the immense influence of all climatic factors during a lengthy part of every annual vegetation period. In this situation, the basic trends outlined, take on the character of properties, i.e. of constants in a physical sense. A paradigm develops over time that increases in strength and rigidity in the course of time. The basis of this difficulty is that the branches of science that are involved have to change as well. Before the responsible institutions become sufficiently aware of this, it is hard to attract any public attention. Such a situation is developing now within agricultural plant production, when further improvement of "output" in terms of man-hours per unit crop starts to coincide with increasing cases and the amount of collateral damage. Increasing weight and effectiveness of machinery impeding water movement in the soil profile are such a case.

The frequency of such unwanted events increases. Most important, however, is the fact that for the first time within the whole of technological development in agricultural production, an increase in efficiency of human labour has started to result in damage to the demands of crops. This last tendency is still small but expanding in unexpected places. So neither its development nor the resulting necessity have caused the general paradigm to change. Future development will have to pay attention first to avoiding damage to soil as the primary production factor, and increasing the efficiency of the separate steps will have to take second place (Hartge et al. 2004).

9.2.2 Development or New Aspects Needed

At the time when chemical investigation prevailed, physical factors like soil volume and thus porosity were treated as constant. Volume changes resulting from the preparation of the seedbed were considered as transient and limited in extent. During the last decades of the nineteenth century ploughing became increasingly effective through improvement of the blades. Traction by horses was the only technique that was available at a time when first observations of plough pans (soil shear) below the ploughed topsoil were reported (Schultz-Klinken 1981). Changes of soil surface level and thus of pore volume were considered as short-lived and thus could be neglected. Such pans were understood as obstacles to root penetration into subsoil and thus detrimental for water uptake from subsoil by the roots. The term "plough pan" expressed vividly the opinion about the shaping of this new appearance. These observed processes increased in the course of the development of equipment for seedbed preparation and harvesting. Machines constantly increased in weight, and so did the tractors in order to provide traction for their efficient use. This development implied a slow but regular increase of the difference between vertical and horizontal stresses in the developed profile. But this fact was neither observed nor heeded. The central endeavour of the whole development in soil tillage was to increase the efficiency of human labour and the maximisation of the harvest, in the same way as it had been achieved during the whole course of human technologies since its very beginning.

Considerations of its effect on site properties for plant growth were initially neglected. Fitting the soil water regime had been obtained by choice of the crop. If insufficient crop development was observed, it was considered to be remedied by an increase in fertiliser application. The frequency of observations of decreasing yields, in spite of increasing sophistication in fertilisation technology – mostly observed with cereals – dates only from the last decades of the twentieth century. A general yield depression was reported in the turning-area (headland) at the ends of furrows, where the machinery has to raise its working tools, to turn and to start the next tillage run (Sparkes et al. 1998). But it was concluded that such a state of soil compaction was not likely to develop within the range of furrow and dam if equipment with the most modern technology was being applied. These observations, supported by similar experimental results published occasionally (Heuer et al. 2008), were considered to prove that the machinery used would do no damage to the soil. In general, first evaluations showed that a breaking of the plough pan did enable roots to penetrate deeper soil parts. However, obviously compaction began more and more frequently to reach soil depths that were not affected by routine tillage. For such cases, the new term "subsoil compaction" was coined. When it began to replace the earlier term "plough pan," this at the same time showed a change in general opinion that there was a process increasing in importance which was not a consequence solely of ploughing. The compaction state, in terms of air capacity, saturated water conductivity, and soil bulk density, was measured with core samples taken from soil profiles, and limits set beyond which damage by compaction was considered (Brunotte et al. 2008). All kinds of wheeling at soil water contents, which allow deformation by the applied weights of machinery, in combination with unsuitable tyre profiles and contact pressures – at the soil–wheel interface – were accepted as immediate mechanisms for subsoil compaction. The early stages of this research were characterised by uncertainty about which of the existing methods would be the most suitable to describe the observed changes of soil and their influence on the quality of crop production. This uncertainty became obvious when, forced by the "Iron Curtain", separate methods were developed and discussed in Eastern and Western Europe. The wide-ranging and sometimes contradictory results stimulated more specialised investigations. New terms were created to describe different combinations and single features of the complex "Soil-Compaction-Wheeling-Plant-Root-Water-Regime". The term subsoil compaction became well-known enough to attract specialists (Drescher et al. 1988; Horn et al. 2000). The enormous effort that is necessary to achieve the basic paradigm shift – as explained in the preceding section – is generally still not realised.

The result of meetings encouraged the identification of some points as a basis for further discussion:

1. Soil – like all materials that tolerate plant growth – has no constant volume.
2. Individual solid particles have to retain a minimum mobility to allow permanent root growth. Frictional resistance between grains at each direct contact will add up to give a different resultant movement for each particle. The resultants of mechanical stress acting on each separate grain will not coincide. They will add up to give a resultant in different directions at all solid–solid contacts.
3. Horizontal movement of particles is impeded by neighbouring solid soil particles packed at different levels. Thus the only direction of escape for volume expansion is vertical movement, which means lifting of the solid soil surface.
4. Reliable determination of even minor changes of soil surface levels, in terms of lifting or lowering of soil surface related to some independent reference level, is necessary in order to assess the results of experimental treatments.

9.2.3 Principles Between Cause and Impact

At least two lines have to be worked on to get reliable results of impacts.

Engineering soil mechanics had devised a procedure, which is well known amongst most researchers at different disciplines. It is the "Proctor Test" (Proctor 1948) for obtaining a very high soil bulk density (close to maximum) without massive breakdown of mineral particles, with relatively modest apparatus.

Digression I: The Proctor Test

The Proctor density is a reproducible value when applying the following procedure: A soil sample is subdivided into, say, six subsamples. Each is carefully mixed with water to give different water contents and set aside for thorough equilibration within the sample. The experimental specimen is obtained by placing one third (1/3) of the sample into a tube of 10 cm diameter, and dropping a standardised falling weight (efficient area about one fourth of tube) 25 times. Add the next portion of the sample, repeat the process, add the rest of the sample and proceed as before, then determine dry soil bulk density and water content. If the procedure is followed carefully, a value of the high soil bulk density is obtained with sufficient reproducibility for almost all soil materials, independent of their compound. Repeating the procedure with at least four different values of water content of the sample allows the determination of a maximum at a certain water content.

Strongly specified conditions are needed for different soil materials (Kézdi and László 1980). The time to reach equilibrium depends on particle size distribution, i.e. on texture, on amount and character of organic material in the sample and, last but not least, water content. But at least there is an equilibrium or a state very close to it. Under open air conditions this might never be reached before external conditions change. In Fig. 9.1 an arbitrary sample is shown (left) with an initial void ratio of e.g., one (1). During the process of compaction the solid mass/volume remains constant and so the void ratio decreases (here below one (1)), and inevitably the soil surface decreases. This is the procedure of compaction. Under normal on-site conditions, this height loss will not be seen because the amount may be very small and there is no reference level. If a soil layer has been loosened by tillage, e.g. ploughing, its total volume increases. As there is no chance to expand sidewise, volume increase can only be obtained by lifting up of the soil surface (Fig. 9.1, right).

At this moment, another important point joins the discussion of soil density and compaction problems: the very widely applied term "porosity". It is a vivid one and can be directly determined. However for this kind of problem the term is inappropriate as a reference for changes because it is based on total volume in contrast to the "void ratio". The term "void ratio" is much less commonly used in agricultural soil science. It is not as vivid as "porosity" and therefore frequently avoided

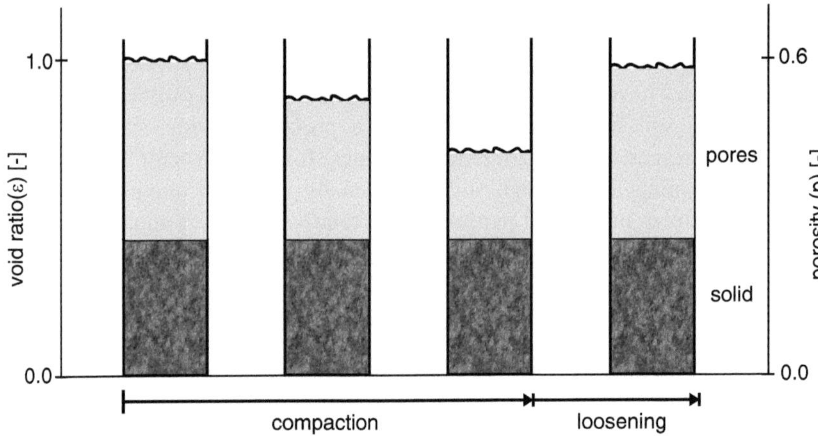

Fig. 9.1 Changes of soil surface level connected in relation to the change in porosity (*right*) and void ratio (*left*)

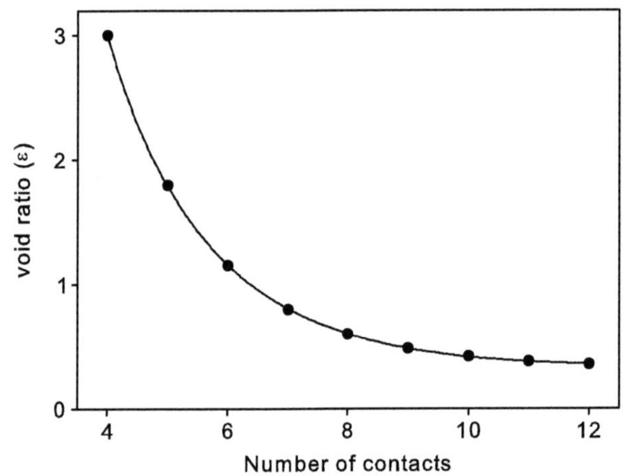

Fig. 9.2 Void ratio in relation to the number of solid–solid contacts per solid particle

in favour of "porosity". As shown in Fig. 9.1, the constant solid volume is the reference basis. For the discussion of processes which include volume changes such as the change in the level of the soil surface, "void ratio" is the more appropriate term.

Still another relation should be kept in mind because it explains the reaction of grainy material to changes in environmental situation. Any change in soil surface level is accompanied by a change in the number of solid–solid contacts per individual solid particle. Although the relation as shown in Fig. 9.2 is valid in its strictest form only, when the material consists of perfect spherical grains of equal size, it does however depict clearly the importance of the independent general relation and demonstrates clearly the general reaction of the soil to pressure. It is obvious that any mineral grain becomes less mobile when the number of contacts with its neighbours increases. Expressed the other way round, this means that increase of stability of the structure of any soil is unavoidably correlated with an increasing number of contacts. This seems to be in contradiction to the frequently observed complete loss of stability when wet soil is trafficked. It is important at this point to remember that this loss of strength is a consequence of a drastically changed stress situation of the water in the topsoil if drainage is impeded. This however is again the first consequence of deformation, which means that the first reaction to any increasing load is increasing contacts where they are at a minimum. Such places are the mineral grains at the fringes of large pores which would have afforded quick draining of infiltration water during the precipitation. Another consequence of this mechanism is a change in the whole water regime of the site. Periods of wet, weak soil will become more frequent and last longer; most of the large pores are destroyed because of inhibited infiltration. Further decreasing infiltration means more evaporation and, with this, further shrinkage of the whole mass of the particular aggregates. What are the forces that produce these mostly only small movements of the soil surface? Since freshly accumulated soil material settles downwards, gravitation is the most obvious active force. Close observation however shows which are the main causes which can lead to subsoil compaction:

1. Water impact – a natural cause which can create thin soil crusts that can delay or prevent seedling emergence. Further reasons can be the change in the menisci forces within the soil, or the water's own weight.
2. Tillage operations – the intensive or continuous working and ploughing within the same depth range causes extensive plough pans and forms compacted subsoil layers.
3. Impact of any temporary machinery load – as mentioned before this is the main reason for subsoil compaction due to the aforementioned increase in field and farm sizes and the increasing weight of agricultural machinery correlated with the decrease in manpower in modern agriculture.
4. Minimisation of crop rotation – the trend of declining crop rotation can cause effects such as limited rooting systems, which may increase subsoil compaction, or the increasing possibility of compaction in the early cropping season due to higher tillage and field traffic activity.

Following these kinds of combinations, it does not seem surprising that a rather lengthy procedure is necessary to obtain the Proctor density. In agricultural practice, a "real equilibrium" is most difficult to observe and is remote at any arbitrary moment. Upward movement of the soil surface is obvious to everybody who walks attentively on, e.g., a paved footway, a trampled footpath, a sports field, a grazed meadow (pasture), a field for hay-making and finally a footpath in a forest. Any softening of the soil surface is caused, as explained earlier, by heaving up of soil material and thus decreasing numbers of solid contacts per grain by biological processes such as root growth or animal activity (burrowing) and, in special situations, the growth of ice crystals.

9.3 Impacts

Subsoil compaction can lead to rapid, intense, longlasting, irreversible and persistent changes in the soil structure characteristics and soil transport functions. The change in the pore system and the pore geometry in a compacted subsoil turns from a vertical pore system into a horizontal pore system with a primarily platey structure (Horn et al. 1995). This platey structure has the attribute of a boundary layer, which primarily limits water and air conductivity and storage capacity, the oxygen supply of plant roots, plant root growth and soil life. These changes can cause various and extensive impacts which can be differentiated by origin and by magnitude of the effects. The first and direct impacts of subsoil compaction for the farmer can be the delay of planting or other field operations due to colder (change of the soil heat balance) and wetter soils (change of the soil water storage capacity). For example, the changed heat balance means that compacted soils heat up and cool down more slowly, so that frost heaving as a natural amelioration process can be excluded. Furthermore, the higher packing density can result in greater complexity (machinery, fuel) for ploughing compacted soil. The extensive deformation of soil aggregates and the higher bulk density increases the soil strength, which is accompanied with higher penetration resistances. The higher penetration resistance is a limiting factor for rootability and root growth (root deformation) which can result in a lack of water and nutrient supply and cause poorer plant growth and higher vulnerability of the crop to diseases. Soil life can be affected by subsoil compaction in such a way that the population density and the species spectrum of soil life and microorganisms change (an example is the change in earthworm species and their population density), while biological activity is limited and resultant biochemical processes become increasingly anaerobic. Hence, the soil fertility will be reduced. Subsoil compaction and the reduced water infiltration capacity can result in an intense water storage between soil surface and subsoil whereby the water persists significantly longer than normal rainfall surface water. Besides this, the high soil water storage induces massive reduction of aeration, which can lead to anaerobic conditions. This too can reduce plant growth, make the crops more vulnerable to diseases and foul or limit the nutrient supply (e.g. an intense loss of N and K uptake) (Arvidsson 1999). Worst-case anaerobic conditions caused by subsoil compaction can lead to a production of greenhouse gases (N_2O, CH_4) (Ruser et al. 1998). Another far-reaching effect of subsoil compaction is the contribution to floods. Either the infiltration rate in compacted soil is too low, or intense or long-term precipitation is completely unable to infiltrate into the subsoil and deeper layers. This can result in surface run-off or lateral run-off into the next aquifer and locally to higher floods and flood intensities. Research has shown that the change in agricultural land use can not only lead to floods during high precipitation periods but also particularly into flood magnification at normal precipitation amounts (Pinter et al. 2006; van der Ploeg et al. 2002). The lack of infiltration capacity can also favour a higher erodibility or erosion and so a considerable loss of soil material. Similarly, the higher transport amount and velocity of organic waste and agrochemicals into the recipient area can cause massive damage to biotope systems and surface water (eutrophication). All these impacts of subsoil compaction can cause direct yield depressions. Besides this, there are higher costs in agricultural management, e.g. intensive machinery use, higher consumption of fuel or agrochemicals, and additionally high follow-up costs to avoid further or remedy present consequences of the subsoil compaction on a local and global scale, e.g. re-establishment, flood magnification, erosion effects and climate change effects.

9.4 Detection and Identification

9.4.1 Visual Detection

9.4.1.1 Visual in-situ Observation

Visual in-situ observation is probably the easiest way to detect compaction. Generally, the farmer will have tilled his field most carefully to obtain even germination and crop development until harvest in order to guarantee uniform quality of his product. If therefore differences in plant development, i.e. growth, flowering time, and degree of ripening, become visible (Fig. 9.3), he will suspect disturbances in the local water regime. Each farmer will know local areas on every plot where the disturbances will develop during the annual weather regime and observe the

Fig. 9.3 Visible impeded plant development by non-uniform traffic lanes

development over the course of years. Today the first "suspect" for disturbance of plant growth will be related to soil compaction and this is more probable the heavier (i.e., newer) the machinery used. Spade pits or profile pits are further in-situ opportunities to visually access subsoil compaction and its effects on, e.g., soil rootability, aeration status, and drainability. In those pits, differences in the stratification, but more likely differences in the soil structure, can be observed, e.g., subsoil compacted areas in loess soils are mainly related to impermeable plate/platey structures. At least two pits are necessary for evaluation: one pit in the "suspicious" and one pit in the "non-suspicious" area to see if the observed differences have no other (e.g., geological) origin. But without any measurable data, these observations are just (unreliable) visual information.

9.4.1.2 Remote Sensing

Another possibility for visual observation can be remote sensing. Aerial photogrammetry or satellite imagery can be suitable tools for the detection of subsoil compacted areas. This visual information, provided for a larger scale, can be helpful in identifying regional differences in, e.g., growth or flowering. Dependent on the resolution, the results can give direct or indirect evidence of the abovementioned visual subsoil compaction indicators. However, this kind of observation needs subject-related expertise. Figure 9.4 shows such a situation,

Fig. 9.4 Air photography of decreased plant development in rapeseed due to waterlogging

where direction and length of track-ruts show how pedological wetness areas are enlarged by wheeling. In the case shown here, it is not so much wetness but more the slow but inevitable increase of particular situations which needs further attention with regard to compaction. The intensive use of remote sensing tools is mainly used where precision agriculture technologies are in place. The disadvantage is that this visual information can give no detailed information as to whether the observations have different origins or not. To confirm the suspicion of subsoil compaction, various invasive and labour-intensive methods still have to be put into operation.

9.4.2 Invasive Measured Data

9.4.2.1 Vertical Soil Sensors

Whereas visual assessment in most cases needs aerial observation, soil softness can be identified on-site. Simple methods are most useful, and may provide sufficient data on local conditions to draw a map of the local status. Measurement of the soil strength should be considered primarily. Here, vertical penetration resistance measured with simple hand-driven probes (penetrometer) is the first option (Bachmann et al. 2006). These probes (e.g., a digital penetrometer such as the Penetrologger by Eijkelkamp) permit, in relation to the soil pits or soil core sampling (Sect. 9.4.2.3), an in-situ analysis of the compaction state of the soil up to 0.80 m (digital penetrometer) or up to 1.5 m depth (analogue mechanical penetrometer). The measurements are time-saving because they can be done quickly by a single person with, e.g., about 400 penetrations a day, with results immediately available. Additionally, the soil damage is only small (minimal-invasive). The disadvantage is the distinct dependency on the soil water content. Penetrologger measurements cannot determine the soil's stress situation at all, which is be essential for the interpretation of subsoil compaction. Using the Penetrologger, it is evident that the input force required to push the penetrometer into the soil can be higher in one field area than in another. Less labour-intensive alternatives can be hydraulically powered penetrometers, such as a tractor-mounted multi-penetrometer (Domsch et al. 2006), where four penetrometers measure the penetration resistance (PR) in parallel (max. depth 0.6 m) or a Veris 3000 Profiler which measures the PR and the electrical conductivity with a single probe mounted on a cart mast (max. depth 0.9 m) (Sudduth et al. 2004). No matter what probe type is used, a general problem for the question of scaling can be the high spatial variability of the PR results. Large probing intervals (e.g., 1 m distance) can cause a very low spatial dependency (Domsch et al 2006). For further interpretation, knowledge of the geological situation is indispensable. Nevertheless, the Penetrologger is a helpful tool for pre-screening, e.g., to locate the right spot for soil pits or soil core sampling.

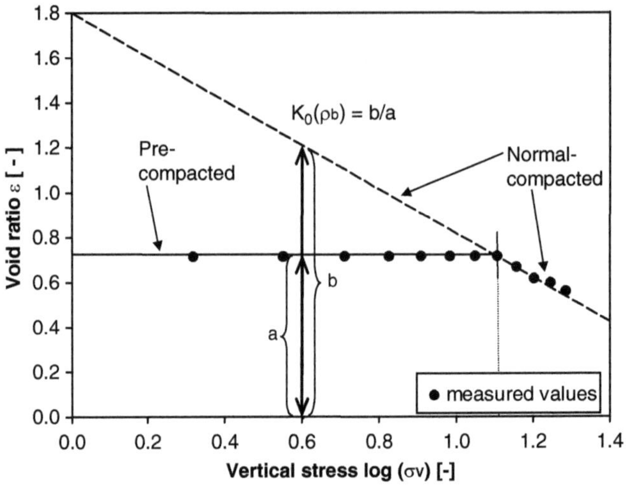

Fig. 9.5 Determination of the stress-at-rest coefficient K_0 by using depth-dependent soil bulk density data; a = pre-compaction (PC), b = normal-compaction (NC)

Digression II: The Reference System

As described in Sect. 9.4.2.1, a measurement of soil strength with a hand-driven probe like the penetrometer can be considered as a simple tool to identify local soil stress differences and may provide adequate data to construct a map of the value distribution of the site. In order to obtain results which are comparable and independent of soil parameters like texture and local water content, a procedure was worked out which ranks the measured data with regard to soil depth and stress distribution. As the reference level for this, the coefficient of stresses at rest (K_0) was chosen as defined in constructional soil mechanics (Hartge 2001; Kezdi 1980). Further reports (Bachmann and Hartge 2006) show, that an observed similarity of readings indicates that the horizontal stress component is dominant for vertical penetration resistance as well as for shear resistance. Readings from both measurements may be used to represent the horizontal stress component in order to estimate an equivalent of the stress-at-rest coefficient K_0, where K_0 is the ratio of normal-compaction and pre-compaction (Fig. 9.5). The state of pre-compaction can be detected using results of PR measurements. The underlying assumption is that the vertical stress component for the lowest layer assessed by PR measurements represents the hydrostatic stress situation, i.e., the stress in that depth is uniform in all directions. The procedure described by Hartge and Bachmann (2004) proposed that drawing a straight line from the maximum depth towards the origin in the depth versus PR plot gives values of the hydrostatic state at each depth for a mechanically non-affected soil – i.e., values for the principal stress (σ_x) are available for each depth up to the soil surface simply by linear interpolation (Fig. 9.6). Deviations from the ideal (hydrostatic) condition, which serves as the reference for non-compaction, are considered to represent the depth-dependent compaction state of the soil, i.e. K_0 values > 1 indicate compacted soil layers and K_0 values < 1 represent loosened layers (Horn et al. 2007).

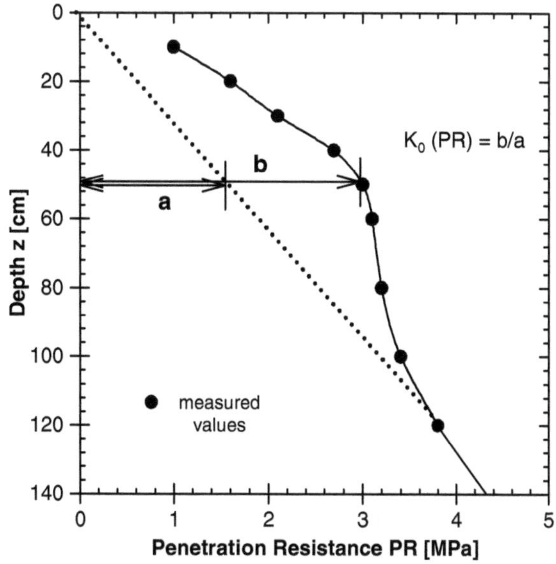

Fig. 9.6 Determination of the stress-at-rest coefficient K_0 by using vertical cone penetration data

9.4.2.2 Horizontal Soil Sensors

In an attempt to construct less labour-intensive machinery to assess as much information as possible about soil compaction, tractor-pulled force sensors have been developed. These soil strength sensors are moved horizontally through the soil while registering the resistance force from cutting, breakage or displacement of the soil. Primarily developed to assess near-surface soil compaction, these tools can also be used for the subsoil area (>25 cm). With the concept of a relationship between horizontal and vertical soil stresses, comparable results of the two types of penetrometers can be expected. The sensors are mainly tip-based or tine-based and are combinations of blades or subsoiler chisels with parallel soil water content sensors. Cone- or prism-shaped penetrometer tips are mounted on specially designed shanks or blades and coupled with load cells which, in combination with a force lever, function as a force sensor (Zeng et al. 2008; Hemmat and Adamchuk 2008). They can work as single- or multiple-tip sensors to assess either one depth or different depths simultaneously. Results of field experiments show that the results of horizontal penetration resistance measurements are comparable to vertical measurements with statistical significance, especially in depths below 25 cm, but not above this depth (Hemmat and Adamchuk 2008). These kinds of sensors have the advantage that they can easily be used for mapping horizontal soil strength and soil water content on a larger scale and therefore can indicate subsoil compaction (Sudduth et al. 2008). But like the measurements with the vertical penetrometer, their dependency on soil water content can limit the reliability of the results, which are valid only within a small range around field capacity.

Additionally, the results might only be limited to one measurement event, due to the cutting or breakage of the soil and depending on the specific sensor-to-sensor distances. Therefore reproducibility or verifiability by invasive methods, from e.g., soil pits, might not be possible.

9.4.2.3 Soil and Soil Core Sampling

Probably the most practical way to assess the change in soil rootability, aeration status and drainability is the time-consuming and labour-intensive but very accurate method of digging a profile pit. This method permits a very detailed depth-related view of the conditions in the profile, especially of root growth, soil structure and soil pore system and therefore of the soil stability and the drainage system. To substantiate the first visual impressions (Sect. 9.4.1.1), soil samples are taken for detailed analysis in the laboratory, which includes the grain size distribution as a tool for the compaction risk assessment. But more important are soil core sample rings to analyse, e.g., bulk density, hydraulic conductivity, aerial conductivity and permeability, compression, shear stress, rheology, and the precompression stress (Hartge and Horn 2009). The analysis of these parameters and the knowledge gained thereby can lead to an exact depth-resolved prediction of whether the suspicious area is affected by subsoil compaction or not. The advantage of these methodologies is the exact and detailed results. The disadvantage, on the other hand, is that digging soil pits nowadays is too time-consuming, too labour-intensive, and economically ineffective. Furthermore, the results can only be produced in the laboratory and are only valid for the small area examined and cannot usually be transferred to the whole area. Finally, the damage to the soil system and the surroundings is considerable.

9.4.3 Non-invasive Measured Data

9.4.3.1 Correlation with Geophysical Sensors

The alternative to the common labour-intensive methods for the detection of subsoil compaction can be non-invasive geophysically-based methods which are less time-consuming. Geophysical probes allow detection on a larger field scale and are "harmless" to the state of the soil. These methods have their origin in archaeology and geology. Nowadays they can be found in precision agriculture as a discipline called "Agro-Geophysics". The various methods are mainly based on electromagnetic induction (EMI), galvanic constant (electrical) resistivity (GCR), electromagnetic reflection (EMR), magnetometry, and self-potential or seismic measurements (Allred et al. 2008). The methods which can detect various values related to subsoil compaction are mainly EMR probes such as ground penetrating radar (GPR), GCR probes like the "Geophilus Electricus", or EMI probes like the EM38. Common to all methods is the transmission of an electromagnetic pulse into the ground and the recording of the response, which is given, e.g., as the apparent electrical conductivity (EC_a) or the electromagnetic reflection (EMR). An exception is a seismic measurement, where seismic energy is induced to the soil. The velocity of the resulting longitudinally and transversally waves (P- and S-waves) can be related to soil properties. Due to the complex evaluation of the results, the seismic method is not suitable for mapping on a larger scale (Petersen et al. 2005). The most promising probes and their applicability for the detection of subsoil compaction will be described below.

Measurement of Electromagnetic Reflection (EMR)

GPR (Geophysical Survey Systems Incorporation, Salem, USA) uses electromagnetic energy in the microwave band (UHF/VHF frequencies) of the radio spectrum and detects the reflected signals from subsurface structures. The transmitting antenna radiates short pulses – picoseconds or nano-seconds – of the high-frequency waves into the ground. The wave hits, e.g., a boundary layer with different dielectric properties, and the receiving antenna records the reflected return signal. The GPR probably is the most exact technique for a detailed view into the ground without destroying the soil. Subsoil structures can be made visible by interpreting runtime and amplitude. The results may allow a differentiated view of the whole field with a high resolution. The resolution and depth reached below surface depend on the use of the right antenna. To get a higher resolution of the soil structure at smaller depths it is necessary to use a high frequency antenna, e.g., 400 or 900 MHz. The disadvantage of these antennas is that they have to be pulled in parallel lines over the field to get good results and therefore it is time- and labour-consuming. Further detriments are their sensibility to weather-related moisture, metal, any kind of electromagnetic radiation, and especially the rough soil surface which can lead to erroneous response signals. Nevertheless, GPR results coincide quite well with the results of Penetrologger measurements concerning the detection of subsoil compaction (Hoefer et al. 2007a), especially for a depth-related identification of the compacted zones (Hoefer et al. 2009b, 2009c). Therefore, GPR is a good non-invasive tool to get detailed information for smaller areas or sub-areas.

Measurement of Galvanic Constant Resistivity (GCR)

The “Geophilus Electricus” (University of Potsdam and Institute of Vegetable and Ornamental Crops, Großbeeren, Germany) is a galvanic contact resistivity meter which emits an alternating voltage into the ground from rolling metal electrodes and measures the resistance to the flow of the electric current. The system works as a six-electrode-pair array (six channels) – equatorial dipole–dipole array (a Wenner array) – which is pulled behind a tractor. To get better ground contact, the metal coulter electrodes have metal spike extensions. The first electrode pair (channel) function as the current electrode and the additional five electrode pairs (channel) as the potential electrodes with a 1 m dipole width and a 0.5 m dipole spacing. The current electrode pair induces an alternating voltage in the soil while the following electrodes measure the resulting voltage. It is capable of measuring complex conductivity (amplitude and phase shift) in a frequency range between 1 MHz and 1 kHz. Four frequencies and five channels can be measured simultaneously This allows determination of the apparent electrical resistance in five depth sections with an practical maximum depth of 1 m (Lück and Rühlmann 2007, 2008a, 2008b, Lück et al. 2009). The values of the apparent electrical conductivity EC_a are measured in milliSiemens per metre (mS/m). The "Geophilus Electricus" system as a tractor-pulled system allows fast measurements and is especially suitable at larger field scales. The application of this system is time-limited to shallow plant growth to avoid further and future growth disturbances. Nevertheless, the measurements so far show promising results for a depth-resolved detection of subsoil compaction zones and the development of its abilities in detail is progressing (Hoefer et al. 2009a, 2009b).

Measurement of Electromagnetic Induction (EMI)

The compaction state of the subsoil can also be assessed by measuring the apparent electrical conductivity (EC_a) with an EM38 (Hoefer et al. 2005, 2006a, 2006b). The electromagnetic induction meter (EM38 probe; Geonics, Mississauga, Canada) induces an electromagnetic field in the ground with a transmitter coil and the secondary electromagnetic field is measured with a receiver coil, giving the apparent electrical conductivity of the soil. The spacing between the two coils is 1 m. The EM38 reaches, on average, a maximum depth of 1.5 m in the vertical mode and 0.75 m in the horizontal dipole mode. The actual depth depends on the local apparent electrical conductivity of the soil. The operating frequency is about 14.6 kHz and the values are measured in milliSiemens per metre (mS/m). This equipment allows fast recording of the apparent electrical conductivity over the whole area, while the values depend on several soil parameters, mainly water and salt content (Rhoades et al. 1989). Both are strongly correlated with local soil texture and those parameters which are, as stated before, similarly correlated with the parameters of intensity and capacity in the solid soil phase and their interrelation (Hoefer et al. 2007a, 2007b, 2007c, 2008; Sudduth et al. 2003, 2005).

The measured results with non-invasive geophysical methods show values of the reflecting signals which can be drawn, e.g., as EC_a distribution maps (Fig. 9.7). But these results are based on the fact that a subsoil layer which is compacted has in most cases a higher water content and a higher conductivity during the wet season, a hard pan during the dry season and moreover, in general, a higher bulk density. Consequently, the signals of the abovementioned geophysical probes experience a higher signal reflection from compacted zones than from non-compacted zones. Using this knowledge as a basis, Hoefer et al. (2009a) showed

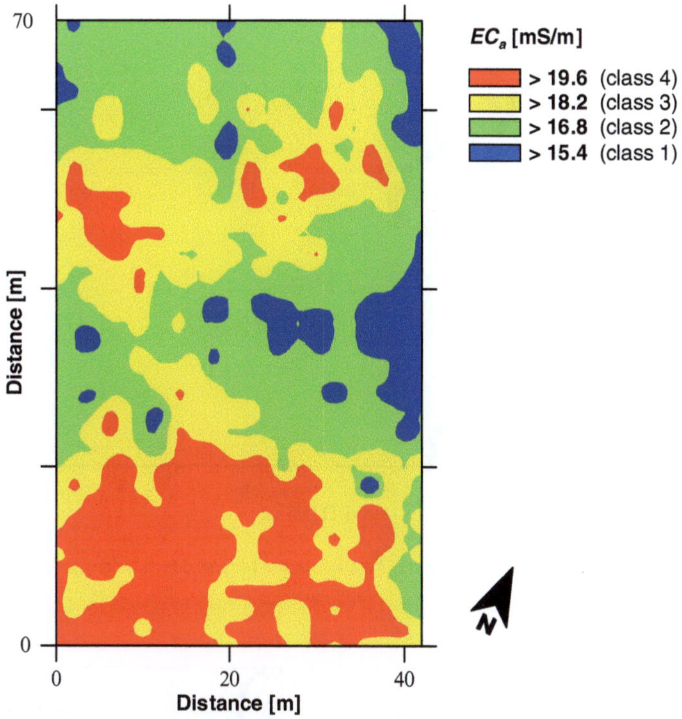

Fig. 9.7 Contour plot of EC_a values (e.g., spring)

that EC_a readings can be correlated with the penetration resistance measurements wherein deviations from a ideal reference line (hydrostatic stress state) are related to higher mechanical stress which results in subsoil compaction (Hoefer et al. 2007c, 2009c). As a result of these assumptions, the area between the measured PR depth function and the hydrostatic state can be expressed as the pre-compaction state of the soil $K_0(PR)$, where $K_0(PR)$ is the sum of the highly compacted area (plough layer) between, e.g., 0.30 and 0.40 m depth (Fig. 9.8). The results show a strong correlation between the Penetrologger (PR) and the responding signal of the geophysical probes, particularly in the depth 0.30–0.40 m, which is generally the depth increment with the highest penetration resistance. A good agreement was also found between EC_a and the pre-compaction state of the subsoil, $K_0(PR)$ (Fig. 9.9). Results show that $K_0(PR)$ is related to EC_a (Fig. 9.10). These principles seem most promising for application to different levels of practical problems because they are fast and time-saving and allow extended evaluations. In general, non-invasive geophysical probes like GPR, the EM38 and the "Geophilus Electricus" are most promising as a pre-screening tool for soil characteristics which are related to subsoil compaction (Hoefer et al. 2009a, 2009b). Development of its facilities in detail is progressing (Hoefer et al. 2009a, 2009b, 2009c).

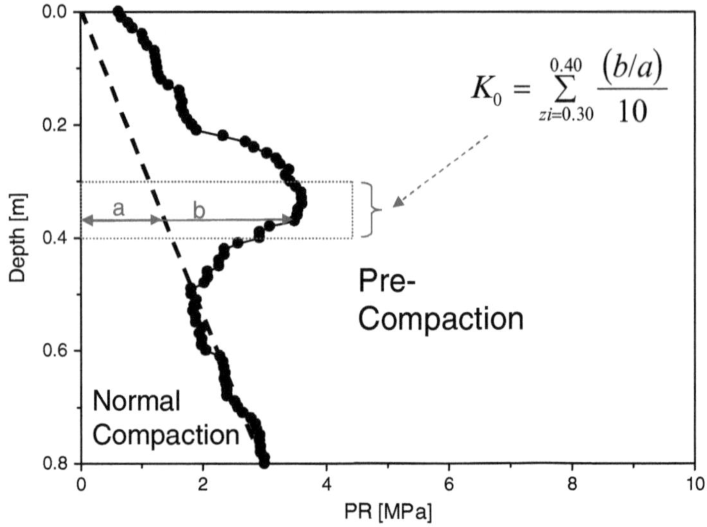

Fig. 9.8 Scheme of parametrisation of the K_0(PR) factor for the depth 0.30–0.40 m

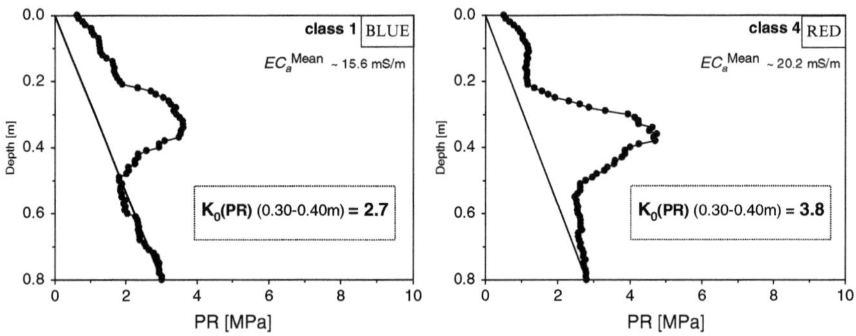

Fig. 9.9 Depth function of penetration resistance (PR) with low values of EC_a (*left*) and high values of EC_a (*right*) (related classes are defined in Fig. 9.7; mean of 10 replicates, e.g., spring)

9.5 Prevention and Re-establishment

9.5.1 Prevention

Within recent decades, the frequency and extent of subsoil compaction in agricultural soils have increased slowly but more or less steadily. There are reports that show some general slowing down of this process as a result of the use of

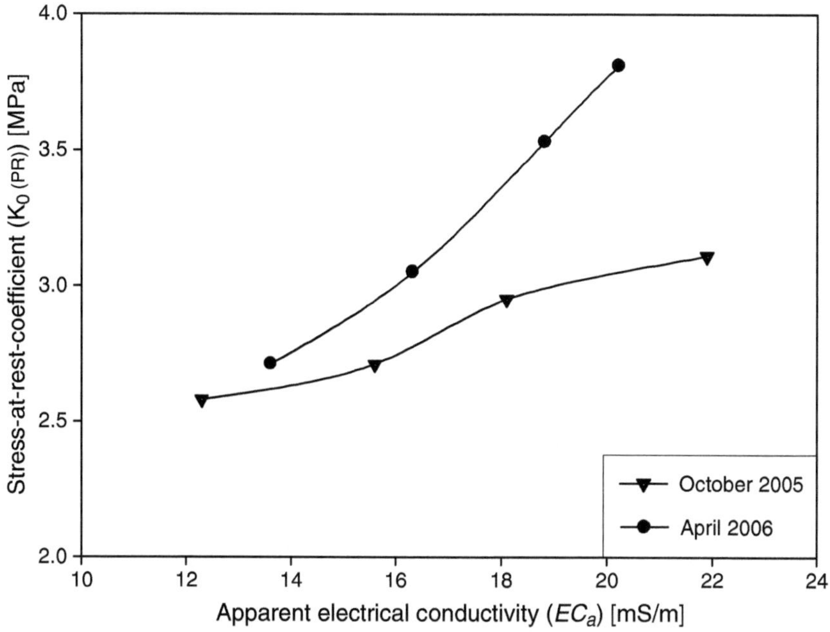

Fig. 9.10 Relationship between K_0(PR) and EC_a (subdivided into four classes)

technologically more sophisticated equipment (Brunotte et al. 2008). The extent of this development is distinct enough to encourage more detailed research. There are certain approaches which deal with the phenomenon as a compound process as well as with its single aspects which result in changes in the strength of soil structure and its stability (Drescher et al. 1988; Horn et al. 2000). These changes became obvious by increasing erosion as well as increasing wetness in cropped areas. An example is the observation in large areas of agricultural cropping in USA where regeneration and increase of field drainage systems is gaining impact. One major step to prevent arable land from subsoil compaction could be the reduction of trafficking on the field, especially taking into account actual weather conditions. Investigations show that usually, during the whole year, an arable field is about 90% traversed with heavy apparatus (Dejong-Hughes et al. 2001). With appropriate and concentrated action – according to good agricultural practices – reduction of trafficking is possible, especially when the actual soil water conditions can lead rapidly to negative effects. Precision agriculture with its management-on-time system could be a solution. Further possibilities could be aligned tillage practices such as on-land ploughing instead of conventional ploughing (off-land), optimisation of the crop rotation system, fertilisation, bed forming, etc. (Chamen et al. 2003). Another approach is reduction of the contact area pressure. By reducing this value the negative depth effects of high tyre pressure – resulting from high vehicle weight – can be decreased to a minimum. To achieve this reduction of contact area pressure, changing to an adjusted tyre type and reducing the tyre inflation pressure are possibilities for minimising the surface deformation, the ground pressure and therefore the effects on subsoil compaction (Chamen et al. 2003; Keller and Arvidsson 2004; Schjønning et al. 2006; Spoor et al. 2003).

A feasible way to reduce the contact area pressure seems to be a switch from smaller to wider tyres, e.g., terra-tyres. Measurements have confirmed it as a good working concept, if the total vehicle weight and axle load is kept constant. Instead the actual development is the construction of more complex and heavier machinery with the aim of reducing the number of vehicle crossings. Due to their complexity and high weight, these vehicles – even with use of terra-tyres – cause deeper and wider pressure bulbs than vehicles with smaller tyres and less weight. Even if the penetration depth of the tyres is reduced (less contact area pressure) and therefore the vertical effect is seemingly less than before, research shows that the problem becomes more and more horizontal. Where the spatial effect of the contact area pressure, e.g., between the two front tyres, was previously limited to the narrow surroundings of the tyre, the effect is now – with terra-tyres – a spatial overlapping of the tyre contact area pressure, especially since the tyre centre has been moved inwardly to avoid making the vehicle too wide (Becher 2005). The main point regarding prevention could only be reduction of the total vehicle weight and thereby a reduction of the axle and the wheel load. All aforementioned means are only minor ways to curtail the already obvious consequences. The origin of the problem will always be too frequent trafficking with too much weight in combination with suboptimal (too wet) soil water conditions.

9.5.2 Amelioration and Re-establishment

Where subsoil compacted layers already exist, amelioration management measures must be considered to reduce the negative effects, especially while biological activity, natural weathering and horticultural measures under most agricultural conditions are too slow or ineffective within normal crop production cycles (Spoor et al. 2003). Knowing that biological, natural, or horticultural re-establishment takes too long, the consequence is mechanical measures as the only solution to accelerate "repair" of the compacted subsoil (Radford et al. 2007). For amelioration of subsoil compaction, the most obvious means is loosening. As pointed out earlier, this would mean lifting up the whole of the compacted zone to break it up and thus increase its porosity. It was understood that this was feasible only in cases where the whole of the profile is of the same geological origin. The result of such a procedure could be judged from the appropriate amount of the soil surface lifted up and be observed easily after the action itself. This action could have existed in deep ploughing and thus result in turning upside down the bulk of the soil, down to about one metre soil depth. Generally, such an intensive change of the whole structure seems to be inconvenient while a tool like a chisel could lift parts of the soil easily. Rigid as well as movable parts were tried. This last technique, called subsoiling, was considered to be less damaging to the soil in case of failure. Both techniques would mean a decrease in the number of contacts per solid particle, as pointed out in detail earlier (Sect. 9.2.). A great number of individual mineral grains would thus be shifted to positions with a small number of contacts, resulting in a destabilising effect. Other options for amelioration can be the creation of fissures or cracks reaching down to the compacted zone, causing only a minimum of disturbance and so optimising at least the root and drainage system (Spoor et al. 2003). This means that the fissuring keeps the loosening of subsoil to a negligible minimum whereas at the same time the macro-pore system is optimised, at least until the next tillage and the following re-compaction. In the 1950s and 1960s, there was a strong momentum towards this kind of amelioration. But observation soon showed that the loosening was mainly short-lived (in the great majority of cases) (Schulte-Karring 1963), which was ascribed to trafficking soon after the amelioration using essentially the same equipment as before. Recent experiments show, that tillage with light trafficking and on-land ploughing results in a more moderate re-compaction of loosened subsoil compared to conventional tillage after subsoil loosening. However, the same research also shows that even with a strong re-compaction the yield depression is just 7% below that of the moderately re-compacted soil, and even the differences in root growth on the yield are negligible (Munkholm et al. 2005a, 2005b). These outcomes seem to be a logical consequence, as a closer look at the mineral particles after amelioration shows. Regardless of the geometric direction of the created planes, an appreciable number of soil particles loosens up one half of its supporting contacts to neighbour grains. In these zones of weakness, reorientation of single grains is the more likely. The particles at "surface" planes of the new aggregates have to travel a longer distance to re-establish sufficient support against still further movements. It is obvious that limiting layers of particles oriented in planes are more exposed to movement in such cases than particles in their original position. Even those at surface tubular or spherical pores or even plain surfaces bent concavely in regard to the main failure of support will retain more support than convex ones. Results obtained with the mentioned techniques in many cases were disappointing. One reason might be that uplifting should move mineral particles exactly in the opposite direction to the compressing strain before amelioration, otherwise re-compaction is most likely at the very first trafficking event after loosening. A new concept is the development and construction of equipment to drill vertical circular tubes instead of forming more or less planes of failure with arbitrary direction by whatever expensive treatment. In golf course or sport field soils such equipment is already used (Dawson 2006). Such tubular pores would be resistant because development of free surface water is unlikely if some secondary action such as seeding covers their open ends (Berkenhagen and Hartge 1996).

9.6 Conclusions

Regarding the historical development from "plough pan" induced by horse-power/lighter machinery to "subsoil compaction" induced by heavier machinery, it seems at first site that subsoil compaction is inevitable with prevailing agricultural methods. However, several approaches show that subsoil compaction is avoidable and because of the far-reaching ecological and economical effects of the local soil state, it must be avoided. Therefore, some prevailing perceptions of the physical properties of subsoil compaction need to be changed. One focus should be on the detection of subsoil compaction. While field or laboratory investigations represent only the local state of soil stress, remote sensing, but especially proximal sensing techniques such as non-invasive geophysical methods, offer great potential for larger scale assessment, and this pre-screening permits concentrated action by farmers to avoid or reduce subsoil compaction. While amelioration or re-establishment actions are unsatisfactory, as well as costly and labour-intensive, and unless future agro-robotics is able to carry out agricultural field work with non-compacting methods, various actions such as reduction of the weight of agricultural machinery, of the contact area pressure, and of the frequency of trafficking are necessary to reduce present and to prevent future subsoil compaction.

Acknowledgement The authors would like to express their special thanks to Prof. Dr. M.B. Kirkham and Dr. S.K. Woche for their professional expertise and also the copyeditor Phil Johnstone for improving the quality of the paper. In addition, we would like to thank the editors Dr. A.P. Dedousis and Dr. T. Bartzanas, as well as Dr. J. Lindenborn from Springer Publishing and Indhu Prasanna from SPi for their patience.

Dedication I dedicate this contribution to my mentor and fatherly friend Karl Heinrich Hartge. He died quite unexpectedly at the age of 83, while working on this text.

References

Alakukku L, Weisskopf P, Chamen WCT, Tijink FGJ, van der Linden JP, Pires S, Sommer C, Spoor G (2003) Prevention strategies for field traffic-induced subsoil compaction: a review: Part 1. Machine/soil interactions. Soil Till Res 73(1–2):145–160

Allred BJ, Daniels JJ, Ehsani MR (eds) (2008) Handbook of agricultural geophysics. CRC Press, Boca Raton, FL

Arvidsson J (1999) Nutrient uptake and growth of barley as affected by soil compaction. Plant Soil 208(1):9–19

Bachmann J, Hartge KH (2006) Estimating soil stress distribution by using depth dependent soil bulk density data. JPNSS 169(2):233–238

Bachmann J, Contreras K, Hartge KH, MacDonald R (2006) Comparision of soil strength data obtained in situ with penetrometer and vane shear test. Soil Till Res 87(1):113–118

Becher HH (2005) Effect of section width, track width and wheel base on the pressure distribution under agricultural vehicles. Mitt Dtsch Bodenkdl Ges 107(1):23–24

Berkenhagen J, Hartge KH (1996) Erosion and accumulation as factors of persistent artificial tube shaped vertical pores in a loess derived luvisol. Z Kulturtech Landentw 37:14–18

Brunotte J, Lorenz M, Sommer C, Harrach T, Schäfer W (2008) The prevalence of harmful soil compaction in southern Lower Saxony. Rep Agric 86(2):262–284

Chamen WCT, Alakukku L, Pires S, Sommer C, Spoor G, Tijink FGJ, Weisskopf P (2003) Prevention strategies for field traffic-induced subsoil compaction: a review: Part 2. Equipment and field practices. Soil Till Res 73(1–2):161–174

Dawson RB (2006) Management of sports turf. Sports Publication, New Dehli

DeJong-Hughes J, Moncrief JF, Voorhees WB, Swan J B (2001) Soil compaction - causes, effects and controls. http://www.extension.umn.edu/distribution/cropsystems/DC3115.html. Accessed on 15 Oct 2008

Domsch H, Ehlert D, Giebel A, Witzke K, Boess J (2006) Evaluation of the soil penetration resistance along a transect to determine the loosening depth. Precision Agric 7(5): 309–326

Drescher J, Horn R, DeBoodt M (ed) (1988) Impact of water and external forces on soil structure. Catena Supplement 11. Catena, Reiskirchen

Hartge KH (2001) Physical state of soil and its evaluation by sophisticated use of simple equipment. R Ciencia del Suelo y N Veg 1(1):41–47

Hartge KH, Bachmann J (2004) Evaluation of the soil consolidation state by using data from penetration resistance probes. JPNSS 167(3):303–308

Hartge KH, Horn R (2009) Die physikalische Untersuchung von Böden. Praxis Messmethoden Auswertung, 4th edn. E. Schweizerbart'sche Verlagsbuchhandlung, Stuttgart

Hartge KH, Horn R, Kretschmer H (2004) Why are the dominating areas of agricultural food production not identical with the initial farming regions? Rep Agric 82(4):479–403

Hemmat A, Adamchuk VI (2008) Sensor systems for measuring soil compaction: review and analysis. Comput Electron Agr 63(2):89–103

Heuer H, Tomanowa O, Koch H-J, Märländer B (2008) Subsoil properties and cereal growth as affected by a single pass of heavy machinery and two tillage systems on a Luvisol. JPNSS 171 (4):580–590

Hoefer G, Bachmann J, Horn R, Schäfer W (2005) Regionalisation of subsoil compaction in agricultural soils as a result of soil compaction induced by heavy machinery trafficking. Mitt Dtsch Bodenkdl Ges 107(1):52–53

Hoefer G, Bachmann J, Hartge KH (2006a) In-situ methods to regionalise subsoil compaction in loess soils under agricultural conditions. Mitt Dtsch Bodenkdl Ges

Hoefer G, Bachmann J, Hartge KH (2006b) Regionalization of subsoil compaction of agricultural soils. In: Proceedings of the 17th Triennial Conference of the International Soil and Tillage Research Organisation (ISTRO), Kiel, Germany

Hoefer G, Bachmann J, Hartge KH, Gundelach V, Altfelder S, Duijnisveld W, Stützel H (2007a) Non-invasive regionalisation of subsoil compaction in loess soils under agricultural conditions. Mitt Dtsch Bodenkdl Ges 110:97–98

Hoefer G, Bachmann J, Hartge KH, Gundelach V, Altfelder S, Noell U, Duijnisveld W (2007b) Detection, regionalisation and classification of subsoil compaction in agricultural loess soils by electromagnetic induction probing. In: Proceedings of the 7th Biennial International Conference on Pedometrics, Tübungen, Germany

Hoefer G, Bachmann J, Hartge KH, Stützel H, Gundelach V, Altfelder S, Noell U, Boess J, Duijnisveld W (2007c) Detection of subsoil compaction in agricultural loess soils by electromagnetic induction probing. In: Proceedings of the ASA-CSSA-SSSA International Annual Meeting, New Orleans, Louisiana, USA

Hoefer G, Bachmann J, Hartge KH (2008) Is a detection and regionalisation of subsoil compaction with an EM38 probe possible? In: Proceedings of the 1st Global Workshop of High Resolution Soil Sensing and Digital Mapping. Sydney, New South Wales, Australia

Hoefer G, Gundelach V, Lück E, Rühlmann J, Bachmann J, Altfelder S (2009a) Proof of subsoil compaction via geophysical probing methods.In: Proceedings of the 69th Annual Meeting of the German Geophysical Society (DGG), Kiel, Germany

Hoefer G, Lück E, Gundelach V, Rühlmann J, Bachmann J (2009b) The capability of non-destructive geophysical methods in precision agriculture to capture subsoil mechanical strength. In: van Henten EJ, Goense D, Lokhorst C (ed) Precision agriculture '09. Proceedings of the 7th European Conference on Precision Agriculture, Wageningen, the Netherlands, pp 439–444

Hoefer G, Lück E, Rühlmann J, Bachmann J (2009c) Flächenhafte Ermittlung der Bodenspannungsänderung durch geophysikalische Sensorik. In: Proceedings of the Biennial Meeting of the German Soil Science Society (DBG), Bonn, Germany

Horn R, Domzzal H, Slowinska-Jurkiewicz A, van Ouwerkerk C (1995) Soil compaction processes and their effects on the structure of arable soils and the environment. Soil Till Res 35(1–2):23–36

Horn R, van den Akker JH, Arvidsson J (ed) (2000) Subsoil compaction, distribution processes and consequences. Advances in Geoecology 32. Catena, Reiskirchen

Horn R, Hartge KH, Bachmann J, Kirkham MB (2007) Mechanical stresses in soils assessed from bulk-density and penetration-resistance data sets. Soil Sci Soc Am J 71(5):1455–1459

Keller T, Arvidsson J (2004) Technical solutions to reduce the risk of subsoil compaction: effects of dual wheels, tandem wheels and tyre inflation pressure on stress propagation in soil. Soil Till Res 79(2):191–205

Kézdi Á, László R (1980) Handbook of soil mechanics: Soil testing, Vol 2. Elsevier, Amsterdam

Lück E, Rühlmann J (2007) Geophilus electricus – a new soil mapping system. Mitt Dtsch Bodenkdl Ges 110:607–608

Lück E, Rühlmann J (2008a) Soil mapping with the Geophilus. In: Proceedings of the 68th Annual Meeting of the German Geophysical Society (DGG), Freiberg, Germany

Lück E, Rühlmann J (2008b) Electrical conductivity mapping with Geophilus Electricus. In: Proceedings of the 14th European Meeting of Environmental and Engineering Geophysics, Krakow, Poland

Lück E, Spangenberg U, Rühlmann J (2009) Comparison of different EC-mapping sensors. In: van Henten EJ, Goense D, Lokhorst C (ed) Precision agriculture '09. Proceedings of the 7th European Conference on Precision Agriculture, Wageningen, the Netherlands, pp 445–452

Munkholm LJ, Schjønning P, Rüegg K (2005a) Mitigation of subsoil recompaction by light traffic and on-land ploughing: I. Soil response. Soil Till Res 80(1–2):149–158

Munkholm LJ, Schjønning P, Jørgensen MH, Thorup-Kristensen K (2005b) Mitigation of subsoil recompaction by light traffic and on-land ploughing: II Root and yield response. Soil Till Res 80(1–2):159–170

Petersen H, Fleige H, Rabbel W, Horn R (2005) Applicability of geophysical prospecting methods for mapping of soil compaction and variability of soil texture on farm land. JPNSS 168(1): 68–79

Pinter N, van der Ploeg RR, Schweigert P, Hoefer G (2006) Flood magnification on the River Rhine. Hydrol Process 20:147–164

Proctor RR (1948) Articles on soil compaction. In: Proceedings of the 2nd Intern Conf Soil Mech and Foundation. Eng.Bd.5, Rotterdam

Radford BJ, Yule DF, McGarry D, Playford C (2007) Amelioration of soil compaction can take 5 years on a Vertisol under no till in the semi-arid subtropics. Soil Till Res 97(2):249–255

Rhoades JD, Lesch SM, Shouse PJ, Alves WJ (1989) New calibrations for determining soil electrical-conductivity – depth relations from electromagnetic measurements. Soil Sci Soc Am J 53(1):74–79

Ruser R, Flessa H, Schilling R, Steindl H, Beese F (1998) Soil compaction and fertilization effects on nitrous oxide and methane fluxes in potato fields. Soil Sci Soc Am J 62(6):1587–1595

Schjønning P, Lamandé M, Frede AT, Arvidsson J, Keller T (2006) Distribution of vertical stress at the soil-tyre interface: effects of tyre inflation pressure and the impact on stress propagation in the soil profile. In: Horn R, Fleige H, Peth S, Peng X (eds) Soil management for sustainability. Advances in Geoecology 38. Catena Verlag, Reiskirchen, pp 38–46

Schulte-Karring H (1963) Amelioration of compacted soils by subsurface loosening and deep manuring. Landw Forsch 17:40–48

Schultz-Klinken K-R (1981) Haken, Pflug und Ackerbau. Ackerbausysteme des Saatfurchen- und Saatbettbaues in urgeschichtlicher und geschichtlicher Zeit sowie ihr Einfluß auf die

Bodenentwicklung. Schriftenreihe für das Deutsche Landwirtschaftsmuseum 1. August Lax Verlagsbuchhandlung, Hildesheim

Sekera F (1951) Gesunder und kranker Boden. Parey, Berlin

Sparkes DL, Jaggard KW, Ramsden SJ, Scott RK (1998) The effect of field margins on the yield of sugar beet and cereal crops. Ann Appl Biol 132(1):129–142

Spoor G, Tijink FGJ, Weisskopf P (2003) Subsoil compaction: risk, avoidance, identification and alleviation. Soil Till Res 73(1–2):175–182

Sudduth KA, Kitchen NR, Bollero GA, Bullock DG, Wiebold WJ (2003) Comparison of electromagnetic induction and direct sensing of soil electrical conductivity. Agron J 95(3):472–482

Sudduth KA, Hummel JW, Drummond ST (2004) Comparison of the Veris Profiler 3000 to an ASAE-standard penetrometer. Appl Eng Agric 20(5):535–541

Sudduth KA, Kitchen NR, Wiebold WJ, Batchelor WD, Bollero GA, Bullock DG, Clay DE, Palm HL, Pierce FJ, Schuler RT, Thelen KD (2005) Relating apparent electrical conductivity to soil properties across the north-central USA. Comput Electron Agr 46(1–3):263–283

Sudduth KA, Chung SO, Andrade-Sanchezc P, Upadhyayad SK (2008) Field comparison of two prototype soil strength profile sensors. Comput Electron Agr 61(1):20–31

van der Ploeg RR, Machulla G, Hermsmeyer D, Ilsemann J, Gieska M, Bachmann J (2002) Changes in land use and the growing number of flash floods in Germany. In: Steenvorden JHAM, Claessen FAM, Willems J (eds.) Agricultural effects on ground and surface waters: research at the edge of science and society. IAHS-Publication 273. IAHS-Press, Wallingford, pp 317–321

Zeng Q, Sun Y, Schulze Lammers P, Ma D, Lin J, Hueging H (2008) Improvement of a dual-sensor horizontal penetrometer by incorporating an EC sensor. Comput Electron Agr 64 (2):333–337

Chapter 10
Testing of Soil Tillage Machinery

Daniele Pochi and Roberto Fanigliulo

10.1 Aims of the Tests

Tests on soil tillage machinery aim at studying the behaviour of the machines from the point of view of their dynamic and energetic performances and of the quality of work. If correctly obtained, the results of the tests provide a significant contribution to knowledge of the machines: it is possible to evaluate prototypes of and modifications to existing machines; the performances of different machines can be compared; it is possible to determine the best coupling between operating machine and tractor in order to ensure quality of work and energy saving, etc. (Perfect et al. 1997; Perdock and Kouwenhoven 1994). The testing should represent an instrument that the manufacturers can use for development of their products and for improvement of their quality. A standardized testing, systematically applied in order to certificate machine performances, should also represent a warranty for the users/farmers that have to choose a machine.

10.2 Classification of the Machines

From the point of view of the conditions of the soils on which they must be used, two main categories of machines are considered.

Machines operating on untilled soil:

- Primary tillage of the seed-bed, such as spike tooth scarifiers, angular tooth scarifiers, claw tooth scarifiers, and vibrating scarifiers;

D. Pochi (✉) and R. Fanigliulo
CRA-ING Agricultural Research Council, Agricultural Engineering Research Unit, Via della Pascolare 16, 00016 Monterotondo (Rome), Italy
e-mail: daniele.pochi@entecra.it; roberto.fanigliulo@entecra.it

A.P. Dedousis and T. Bartzanas (eds.), *Soil Engineering*, Soil Biology 20,
DOI 10.1007/978-3-642-03681-1_10, © Springer-Verlag Berlin Heidelberg 2010

- Soil inverting, such as mouldboard ploughs, disk ploughs, rotary ploughs, and diggers;
- Soil inverting and ploughing, such as mouldboard subsoil harrows, subsoil disk ploughs, and subsoil diggers.

Machines operating on tilled soil:

- Secondary tillage of the soil, such as spring tine cultivators, rigid tine cultivators, rotary cultivators, spring tooth harrows, spike tooth harrows, disk harrows, radial harrows, powered rotary harrows, tandem harrows, rotary harrows, combination harrows, clod crusher rollers, furrow rollers, and rotary hoes;
- Soil compacting, such as packer rollers, clod crusher rollers, and furrow rollers;
- Weed working and mechanical control, such as rake harrows, tooth weeders, spring tooth weeders, disk weeders, serrated disk weeders, and multi-mill rotary weeders;
- Soil packing, such as disk packers, rotary packers, and turn-furrow packers.

The operating machines can also be divided into passive machines (simply pulled by the tractor) and active machines, driven by tractor power take-off.

10.3 Test Methodology

Tests on soil tillage machine performances are usually carried out applying methodologies in compliance with the requirements of consolidated standards (ISO, EN, UNI, ENAMA). Because of the complexity and heterogeneity of the parameters to be determined (Peruzzi et al. 1999), different test methods have been developed referring to different measurement categories, as follows.

10.3.1 Characteristics of the Machines and their Correct Adjustment

An accurate description of the characteristics (dimensions, mass, etc.) and working method of the machine to be tested should be provided. Considering the test conditions and the type of operation for which it is destined, it must be suitably adjusted, according to the indications of the manufacturer, through the execution of preliminary tests aimed at finding, for instance, the best conditions of working velocity, working depth and width, power take-off speed, etc. An important factor in the execution of the tests is represented by the tractor, the dimensions of which (power and mass) must be proportional to the tested machine. The tractors used in the tests should be of good efficiency; in order to evaluate the tractor–operating machine system from the point of view of power requirements, fuel consumption, etc., tractor performances should be known, for instance, through the availability of the actual characteristic curves of the engine that depict the reference conditions with which the field data can be compared. The tractor–implement system should also be equipped with instruments capable of measuring the main parameters during the work, such as digital encoders (velocity and slip), fuel consumption meter, load cells and/or torquemeter, etc.

10.3.2 Test Conditions

The test conditions must be representative of the normal use of the machine to be tested. This criterion should guide the choice of the working velocity, depth, width, etc.; each combination of such parameters requires a sufficient number of replications with consistent results (three at least). The tractor is usually used under conditions of four-wheel traction with locked differential. In the case of machines driven by the power take-off, the speed of this must be adjusted (according to the manufacturer's indication), varying the fuel delivery, and the same power take-off speed must be adopted in all the replications.

10.3.3 Determination of the Test Plots and Main Soil Characteristics

The test plots must be, as far as possible, uniform and have a regular shape. Their length must allow collection of the test data along a reference distance of at least 100 m (base): the pre-defined conditions (of velocity, working depth etc.) must be reached before the beginning of the base and must be kept constant all along it. As to the soil of the test plots, its texture must be determined, according an internationally recognized analytical method (ISSS classification), classifying it as light soil, medium texture soil or heavy soil. The percentage of sand, silt and clay must be recorded. The soil must also be classified by determining, by means of a standard method (BSI 1975), the Atterberg limits (plastic = LP; liquid = LL), that describe its status depending on moisture content. Even if medium texture soils are preferred, all types of soil are capable of providing interesting information on the machines' performances.

10.3.4 Characterization of the Soil

As mentioned above, the machines can be classified depending on their use in untilled or tilled soils. The characteristics of the soil in which the tests are executed must be determined through the measurement of proper parameters.

10.3.4.1 Untilled Soil

Untilled soil and fallow soils must be used for the tests of all the machines performing main tillage as well as preliminary cultivation operations (such as trench ploughs, special scarifiers, etc.). They can also be used for testing machines devoted to soil refinement and to direct preparation of the seed-bed (as in minimum tillage techniques), such as different types of cultivator and harrow, rotary hoes etc. Under such conditions, the main parameters characterizing the soil are: the *moisture content* (%) and the *dry bulk density* (g cm^{-3}) calculated from soil samples extracted at different depths and dried in a oven at 110 °C until the mass becomes constant (Chen et al 1998); the *cone index* (MPa) that expresses the resistance to the penetration in the soil of a standard cone tip as a function of the depth; the *coverage index*, represented by the ratio between the surface covered by biomass (crop residues, weeds, etc.) and the total surface (the methods for the determination of the covered surface range from the counting of the intersections of a standard grid laid on the ground, referred to the total intersections, to image analysis techniques in which, by means of specific software, it is possible to detect the covered areas); the *surface roughness,* intended as the aggregate of rises and dips present on the soil surface and which may range from few millimeters to several tens of centimeters. This is determined on the same soil section, before and after passage of the machine, and is expressed as the standard deviation (σ) of the series of data provided by a manual or electronic profile-meter (Römkens et al. 1988; Raper et al. 2004). The scanning data obtained before and after the tillage will allow calculation of the *roughness index*, σ_r, according to the following relation proposed (Grant et al. 1990):

$$\sigma_r = \left\{ \frac{1}{N-n-1} \sum_{i=a+1}^{N-a} \left[(y_i - \bar{y}_i) - \overline{(y_i - \bar{y}_i)} \right]^2 \right\}^{1/2}, \tag{10.1}$$

where N is the total number or values collected on a pre-defined distance; y_i is the elevation (mm); a = number of values, before and after y_i, used for calculating $\bar{y}_i$ (running mean); $n = 2a + 1$ is the number of values used for calculating $\bar{y}_i$; $\bar{y}_i$ is the mean elevation (mm) referred to progressive intervals the length of which is $M = 2a\Delta x$ (mm), in which Δx is the horizontal distance between two measurements (10 mm). The accuracy of the measurements increases with the number of replications and the average values should be accompanied by the coefficient of variation.

10.3.4.2 Tilled Soil

"Tilled soil" must be a soil ploughed to a depth of at least 25 cm. It is used for tests on secondary tillage, on the preparation of the seed-bed and on consecutive work, such as rake harrows and weeders with crop in cultivation. In this case, capability to control the weeds is estimated by observing the level of crop damage.

In general, most of the parameters described for untilled soils are extended to tilled soil. A further parameter to consider is the *cloddiness*. Sometimes, soil sampling and handling cause a substantial modification of the sample (loose and dry soil, light friable soil) and the determination of cloddiness is scarcely significant. When reliable measurements of the cloddiness are possible, a 50 cm square trench is dug to the working depth. The soil aggregates found in this volume are left air drying for at least 20 min, then they are divided into six size classes by means of hand-operated or powered standard sieves. An *index* (ranging from 0 to 1) is attributed to each class (Sandri et al. 1998). The *cloddiness* is evaluated by observing the percent of each size class mass referred to the sample total mass. Moreover it is also described by the *refinement index* (I_a) calculated as follows:

$$I_a = \sum_{i=1}^{6} \frac{M_i \cdot I_{ai}}{M_t}, \tag{10.2}$$

where $M_i \cdot I_{ai}$ is the product of the mass of each clod size class for the corresponding index (kg); M_t is the total mass of the sample (kg).

For many of the mentioned parameters, the comparison between the values obtained before and after an operation provides further parameters that allow evaluation of the quality of work (see Sect. 10.3.7).

10.3.5 Evaluation of the Operative Performances

The operative performances of soil tillage machines are described by the behavior of a series of parameters that can be directly measured or calculated on the basis of measured data.

10.3.5.1 Working Depth, Tilled Layer, Swarf and Settlement

The *working depth* is the distance between the work bottom and the ground surface mean level (freed of eventual residues). The *thickness of the tilled layer* is the distance between the work bottom and the surface mean level of the tilled soil. The difference between this and the *working depth* represents the *swarf level* in the case of untilled soil, and the *settlement* determined by an implement on tilled soil. These parameters can be measured manually or by means of suitable instruments; the number of measurements must suffice to describe their behavior through the average values and the coefficient of variation, indicating their capability of keeping to the predefined values and the uniformity of the work.

10.3.5.2 Actual and Operating Working Width

The *actual working width* is the width of the strip of land tilled, by the implement, in a single run and measured perpendicularly to the travel direction. The *operating working width* results from the ratio between the total width of a tilled surface and the required number of runs. The ratio between operating working width, b_0, and actual working width, b_a, represents the *working width coefficient* (C_{ww}):

$$C_{ww} = 100\frac{b_0}{b_a}(\%). \tag{10.3}$$

C_{ww} allows evaluation of the runs' superposition and correct estimation of the work capacity.

10.3.5.3 Theoretic and Actual Work Capacity, Working Time, Operating Efficiency

The main parameters derive from time and velocity measurements referred to the test plot length and width, to the machine working width and to the ancillary (turning) time.

The *theoretic work capacity*, C_{tw}, is independent of plot dimensions and turnings, resulting from the product of the actual working width, b_a and the working velocity, v_1:

$$C_{tw} = \frac{b_a \cdot v_1}{10}(\text{ha h}^{-1}). \tag{10.4}$$

The *working time* is referred to the test plot dimensions. The measurements involve the *actual working time*, T_a, and the *turning time*, T_t, the sum of which will provide the *operating time*, T_o, referred to an ideal field, according to the relation:

$$T_o = T_a + T_t(\text{h ha}^{-1}). \tag{10.5}$$

The *operating efficiency* (η_0) of the machine is calculated as the ratio between actual time and operating time:

$$\eta_o = \frac{T_a}{T_o} \tag{10.6}$$

The *operating capacity*, C_{ow}, of the implement is calculated by multiplying the theoretic working capacity by the operating efficiency, as follows:

$$C_{ow} = C_{tw} \cdot \eta_0(\text{ha h}^{-1}). \tag{10.7}$$

10.3.5.4 Fuel Consumption

Fuel consumption depends on the power requirements and losses that occur during the work and the turnings. The measurement can be made manually, by simply filling the fuel tank before starting the test and refilling it after the test by means of a graduated vessel that allows measurement of the volume of fuel needed for restoring the initial level. This determination can be made considering only the work condition (actual consumption) or including the turnings too (operative consumption). The hourly consumption, C_{hv}, results from the ratio between the refill volume, c_v (dm^3), and the running time, T_c, as follows:

$$C_{hv} = \frac{c_v}{T_c}\,(\mathrm{dm^3 h^{-1}}). \tag{10.8}$$

The hourly fuel consumption can be expressed in terms of mass, C_{hm}, multiplying C_{hv} by the diesel density, γ_g (kg dm^{-3}):

$$C_{hm} = C_{hv} \cdot \gamma_g\,(\mathrm{kg\ h^{-1}}). \tag{10.9}$$

The operative fuel consumption for surface unit (of an ideal field), C_{ha}, is calculated as the product of C_{hm} and T_o:

$$C_{ha} = C_{hm} \cdot T_o\,(\mathrm{kg\ ha^{-1}}). \tag{10.10}$$

The fuel consumption can be directly measured by means of proper sensors (flow-meters) installed on the *fuel feed line* of the tractors used in the tests. They generally provide volumetric data that require to be converted to fuel mass consumption, as above described. A further, indirect method can be adopted if updated tractor engine performance data are available: the hourly fuel consumption value can be identified, on the characteristic curve of fuel consumption, through the average value of the engine speed measured during the work.

10.3.5.5 Tractor Slip

Depending on the type of operating machine, tractor slip can be positive in the case of passive implements or negative when a power take-off-driven machine works rotating in concordance with the travel direction, pushing the tractor. The value of the slip is important in the energy balance of the tractor–operating machine system because it allows estimation of the power losses. Minimizing the slip helps in finding the best adjustments of the system. The slip can be easily calculated from the different number of revolutions of a driving wheel on the base both during the execution of the work, n_1, and during the self-dislocation of the tractor–implement system, n_0, under the same conditions of fuel delivery and gear box ratio. Based on such a difference, the slip is provided by the relation:

$$s = 100\frac{(n_1 - n_0)}{n_1}\,(\%). \tag{10.11}$$

10.3.5.6 Force of Traction and Relative Power Requirement

The force of traction is evaluated for passive or mixed (i.e., passive and active). In the case of trailed machines, the force of traction (drawbar pull) can be directly measured by means of a drawbar instrumented with a load cell. More frequently the implements are mounted on the three point linkage and the measurement of the forces requires some specific interventions that will be described below (see Sect. 10.4.2). At any rate, the required parameter is the *net force of traction*, F_{tr}. Multiplying F_{tr} by the working velocity, v_1, provides the *traction power*, W_{tr}, required by the implement:

$$W_{tr} = \frac{v_1 \cdot F_{tr}}{360.1}\,(\mathrm{kW}). \tag{10.12}$$

As well as the net force of traction, it could also be useful to measure the force required for the traction of the tractor–operating machine system at the same working velocity, with gear box in neutral position and implement lifted. This corresponds to the *self-dislocation force*, F_{sd}, and allows calculation of the self-dislocation power, W_{sd}:

$$W_{sd} = \frac{v_1 \cdot F_{sd}}{360.1}\,(\mathrm{kW}). \tag{10.13}$$

W_{tr} and W_{sd} contribute to determining the best coupling between tractor and operating machine.

10.3.5.7 Torque Transmitted by the Power Take-Off and Relative Power Requirements

These are calculated for all soil working machinery having active or mixed actions on the soil. The torque is measured by means of a torquemeter installed between the tractor power take-off and the implement; the product of the torque, M, and the power take-off speed, N_{pto}, provides the *p.t.o. power*, W_{pto}, required by the implement:

$$W_{pto} = \frac{M \times N_{pto}}{955.0206}\,(\mathrm{kW}). \tag{10.14}$$

The self-dislocation power, W_{sd}, can be measured as described in the previous paragraph. W_{pto} and W_{sd} contribute to determine the best coupling between tractor and operating machine.

10.3.5.8 Power Required by the Tractor–Operating Machine System

The total power required by the implement, W_u, can be identified as the sum of the traction power and of the power take-off power requirements:

$$W_{\mathrm{u}} = W_{\mathrm{tr}} + W_{\mathrm{pto}}(\mathrm{kW}). \qquad (10.15)$$

Further components, as hydraulic power, can be sometimes considered if some dynamic function and adjustment of the tested machine is continuously controlled during the work. The ratio between W_{u} and the actual work section provides the *specific power*, W_{sp}, required by the implement:

$$W_{\mathrm{sp}} = W_{\mathrm{u}}\,(b_{\mathrm{o}} \cdot p_{\mathrm{o}})\;(\mathrm{kW\ m^{-1}cm^{-1}}) \qquad (10.16)$$

where b_{o} is the operating working width (m) and p_{o} is the working depth (cm).

As for fuel consumption, if updated characteristic curves of the tractor engine are available and the engine speed during the work has been measured, it is possible to find the total mean power, W_{T}, provided by the engine. If the self-dislocation power, W_{sd}, is also known, it is possible to estimate the power losses for slip, W_{s}, by means of the relation:

$$W_{\mathrm{s}} = s(W_{\mathrm{tr}} + W_{\mathrm{pto}} + W_{\mathrm{sd}})\;(\mathrm{kW}). \qquad (10.17)$$

All these elements contribute to the evaluation of the power balance of the system, suggesting modifications of the operating parameters, with the aim of reducing power requirements and losses and optimizing the coupling between tractor and implement.

Multiplying the tractor total power by the operating working time, T_{o}, allows calculation of the energy required per surface unit:

$$E_{\mathrm{ha}} = 3.6 \cdot W_{\mathrm{t}} \cdot T_{\mathrm{o}}\;(\mathrm{MJ\ ha^{-1}}). \qquad (10.18)$$

Dividing E_{ha} by the working depth, P, gives the energy per unit of volume of tilled soil E_{Vol}:

$$E_{\mathrm{vol}} = \frac{E_{\mathrm{ha}}}{10 \cdot P}\;(\mathrm{kJ\ m^{-3}}). \qquad (10.19)$$

10.3.6 Evaluation of the Quality of Work Performed by Implements That Operate on Untilled Soil

10.3.6.1 Biomass Residues Burying Degree

The degree of burying of the residues, G_i, is calculated from the values of the coverage index determined before and after the implement (see Sect. 10.3.4.1) by means of the relation:

$$G_i = 100\frac{(I_{cs} - I_{cl})}{I_{cs}}(\%), \quad (10.20)$$

where I_{cs} is the coverage index of fallow soil and I_{cl} is the coverage index of tilled soil.

10.3.6.2 Distribution of Biomass Residues in the Tilled Layer

The determination of this parameter requires the digging of trenches (at least three for each passage) allowing observation of the cross-section of the tilled layer. The distribution can be evaluated by means of image analysis techniques applied to pictures of the cross-section: these provide the percentage of surface occupied by residues, referring to 5–10 cm high portions of the cross-section, corresponding to different depths.

10.3.6.3 Other Parameters

Primary tillage determines changes of the characteristics of untilled soil that must be quantified by comparison between the values obtained before and after the work. Characteristics include the dry bulk density, the cone index and the surface roughness determined as described above. A further parameter is the shape (and uniformity) of the work bottom, which should be observed at the same points at which the surface roughness has been measured before and after the tillage. For this purpose, after having accurately indicated the reference points for the measurements, a trench must be dug, transversally to the work direction, reaching the work bottom. This must be cleaned and detected as described for the surface roughness. A final diagram will show the behavior of the surface of the untilled soil, of the soil after the tillage and of the work bottom. The coefficients of variation are calculated from the three series of data, describing the uniformity of the observed profiles and allowing direct comparison of the changes of the surface roughness.

10.3.7 Evaluation of the Quality of Work Performed by Implements That Operate On Tilled Soil

Most parameters that express the changes caused by secondary tillage are determined by comparison of measurements made before and after the execution of the work.

10.3.7.1 Surface Roughness Reduction

The determination of the surface roughness as described in Sect. 10.3.4.1 provides the roughness index before the secondary tillage, σ_{r1} (after the ploughing) and after it, σ_{r2}, at the same point. The degree of roughness reduction, G_{Rr}, as a consequence of the secondary tillage is:

$$G_{Rr} = 100\frac{(\sigma_{r1} - \sigma_{r2})}{\sigma_{r1}}(\%). \quad (10.21)$$

10.3.7.2 Cloddiness Reduction and Seed-Bed Quality

The cloddiness is described by the refinement index (see Sect. 10.3.4.2). The comparison between the refinement index before the secondary tillage, I_{a1}, (after the ploughing) and after it, I_{a2}, provides the *degree of cloddiness* reduction, G_{Cr}, by means of the relation:

$$G_{Cr} = 100\frac{(I_{a2} - I_{a1})}{I_{a2}}(\%). \quad (10.22)$$

The quality of the seed-bed created by secondary tillage is based on the cloddiness values of the soil after the passage of the implement and is expressed by the *seed-bed quality index*:

$$I_q = \frac{M_{\phi \leq 10}}{M_{\phi \geq 10}} \quad (10.23)$$

where $M_{\varnothing \leq 10}$ is the mass of the clods with diameter less or equal to 10 mm and $M_{\varnothing > 10}$ is the mass of the clods with diameter over 10 mm (kg).

10.3.7.3 Weed Control and Crop Damage

These parameters are usually considered in the case of tests on machines for consecutive soil tillage. Weed growth is evaluated by counting the plants present in sample square areas with a 0.5 m side (at least six), reporting the average count value and the CV. The count will provide the number of weeds per square meter before (N_{wb}) and after (N_{wa}) the passage of the implement. The reduction in the number of weeds allows calculation of the *degree of weed control*, G_w, by means of the relation:

$$G_{rm} = 100\frac{(I_{wb} - I_{wa})}{I_{mp}}(\%). \quad (10.24)$$

The crop damage, as a consequence of the passage of an operating machine, is evaluated by counting the number of uprooted, clipped or destroyed plants in sample square areas with a 0.5 m side. The count provides the number of damaged plants per square meter, P_d, that, compared with the plant investment, P_i, allows calculation of the *degree of crop damage*, as follows:

$$G_{dc} = 100 \frac{P_d}{P_i} (\%). \quad (10.25)$$

10.4 Main Instruments and Equipments Used in the Tests

Some of the parameters described in Sect. 10.3 are directly measured by means of instruments and equipment. Scientific and technical developments continuously improve the reliability and accuracy of sensors and data processing systems, considering the severe environmental conditions of farm machinery tests (Ragni and Santoro 1996; Pochi et al. 1996). This allows us to extend the range of tests and to obtain more detailed information on machine performances (Fanigliulo and Pochi 2008). At the same time it requires us to adequately maintain the test system as regards definition of the test objectives, training of the personnel involved in the tests, and periodic calibration of the instruments and their updating. The main instruments and equipment used in the tests have been classified based on the type of measurement they are used for and are described below.

10.4.1 Determination of Soil Characteristics and Quality of Work

10.4.1.1 Core Drill

Soil moisture and dry bulk density must be determined immediately before the tests. Soil samples are collected at depth intervals increasing by 10 cm starting from ground level, usually by means of a drill that supports standard volume steel cylinders (allowing the soil to be introduced in a sufficiently undisturbed form). After drying in the oven (Sect. 10.3.4.1), the soil samples are weighed by means of a precision balance for the determination of dry bulk density and moisture contents.

10.4.1.2 Penetrometer

The penetrometer consists of a probing rod ending in a standard cone top. It is equipped with a load cell and a distance sensor. The resistance the cone encounters when the rod is pushed in the soil and the related depth variations are continuously measured. Soil resistance to penetration can greatly vary, depending on the characteristics of the test plot, requiring a number of measurement sufficient to represent its behavior. Moreover, measurements are affected by such variables as the velocity of penetration, the cone top angle and section, etc. For these reasons, penetrometers usually have standard characteristics and are often provided with a data logger that allow monitoring of the velocity of penetration and storage of a great number of measurements.

10.4.1.3 Soil Profile-Meter

This is used for detecting the profile of soil surface and of the work bottom; from these data, average values of surface roughness, working depth and uniformity of the work bottom can be calculated (see Sect. 10.3.7). Modern profile-meters are commonly represented by *microreliefmeters,* in which laser sensors, moving on an horizontal bar at steps of 1 mm or more, measure the distance from the soil with high accuracy (Fig. 10.1). As the measurements must be made at the same place before and after the tillage, the horizontal bar must exceed the working width: keeping its supports steady allows exact restoration of its position each time.

10.4.2 Determination of Dynamic and Energetic Parameters

10.4.2.1 Load Cells

Load cells are used for the measurement of forces. In soil tillage machinery testing, they are mainly used for determining the force of traction required by the work (Fig. 10.2).

The load cells commonly used are based on an extensimetric principle (variation of an electric signal determined by the deformation of appropriate materials). In order to guarantee the accuracy of measurement, their range of measurement must be proportioned to the type of test: for instance, ploughing tests often require up to 100,000 N full scale load cells whereas for disk harrowing 10,000 N can suffice. As a consequence, a set of load cells with different measurement ranges should be available in a test center. Load cell application modes can be different. In the case of an instrumented drawbar (see Sect. 10.3.5.6), the tractor–operating machine system is pulled by a "traction vehicle" (Fig. 10.3). The *net force of traction* is the difference between the *gross force of traction* (the system is pulled at the same working velocity, with implement working and tractor wheels free), and the force required for the system self-dislocation (system pulled with implement lifted). Alternatively, it is possible to use a suitable frame equipped with a series of load cells that, interposed between the three-point linkage and the implement, is capable of directly measuring all the components of the forces, also providing information on the working of the hydraulic lift.

Fig. 10.1 Laser microreliefmeter used for determining the surface roughness before and after the tillage, the tillage depth and the uniformity of the tillage bottom

Fig. 10.2 Detail of the drawbar equipped with the load cell measuring the force of traction

Fig. 10.3 Dynamometric vehicle during traction tests with a three-furrow plough

10.4.2.2 Torquemeter

Dynamic torquemeters are used for torque measurements at the power take-off (p.t.o.) of the tractor. They usually measure the p.t.o. speed too, allowing direct calculation of the power required by p.t.o.-driven machines. As for the principle of working, the accuracy and the range of measurement, the considerations made for load cells are valid for torquemeters too. They must be suitably attached to the tractor p.t.o., by means of robust steel frames that must perfectly adapt to the shape of the frame surrounding the p.t.o. (Fig. 10.4). Because of the difficulty of these adaptations, test centers usually dedicate one or more tractors exclusively to test activity. An alternative to the torquemeter is represented by the use of a p.t.o. drive shaft equipped with extensimetric sensors capable of detecting the torsion stress that is proportional to the torque applied. As the sensors rotate with the p.t.o. drive shaft, the data must be radio-transmitted to the data logger. The accuracy of this system is lower than for the torquemeter, but if different instrumented p.t.o. drive shafts are available, they can be used with virtually all tractor–implement combinations.

10.4.2.3 Digital Encoders

These sensors measure the number of revolutions made by rotating parts, through the counting of electric pulses: each encoder is characterized by a specific pulse number per revolution (e.g. 100 pulses rev^{-1}); the accuracy of measurement increases with the pulse number per revolution. Dividing the number of revolutions by the time interval during which they have been measured gives the rotation speed (min^{-1}). In the case of a wheel, knowing its under-load radius, the peripheral velocity will be provided by multiplying the rotation speed by its circumference. Digital encoders are installed on the wheels of the tractors, providing their peripheral velocity used in the determination of the slip during work. They are also attached to the p.t.o. when measurements of the engine speed are needed in tests

Fig. 10.4 Torque-tachometer applied at the p.t.o. shaft to measure the torque needed by the operating machine and the p.t.o. speed

that do not require the installation of torquemeters. In such situations the transmission ratio between engine and p.t.o. must be known.

10.4.2.4 Fuel Consumption Measurers

The direct measurement of fuel consumption under dynamic conditions is possible by attaching a volumetric measurer to the fuel feed line. The measuring unit is usually represented by a flow-meter in which the volume of fuel is proportional to the number of pulses counted by a digital encoder. Dividing the volume by the interval of time during which it has been measured provides the fuel flow per unit time. When installing such sensors, it must be that the fuel feed line consists of a *delivery line* and a *recovery line* (for the excess fuel), so that a double measurement is needed. If the instrument is not designed for such a function, a double counter must be installed. At any rate, the fuel consumption will result from the difference between the flow volumes on both lines.

10.4.2.5 Data Acquisition Systems

The sensors described in this section are mounted on different parts of the tractor–implement system. The analog or digital signals describing the behavior of the parameters to be measured are sent to a data logger before being processed by a computer (Al-Janobi 2000; Watts and Longstaff 1989). The data logger is usually lodged on the tractor cab, together with the feed system and eventual signal amplifiers (Fig. 10.5). The data can be stored on memory cards or directly transferred to an on-board PC for immediate processing (McLaughlin et al. 1993). Miniaturization and increasing compactness and reliability of hardware components allow direct acquisition of the data in PCs provided with suitable acquisition cards capable of replacing classic data loggers. Data loggers and acquisition cards must have a sufficient number of channels for the sensors.

The data can be processed after the test, but, thanks to the diffusion of sensors and data processing systems derived from the automotive test sector, real-time test monitoring is becoming frequent in agricultural tests, allowing considerable improvements from the point of view of accuracy of measurement, evaluation of test behavior, reduction of the number of replications, time saving, efficiency of the instruments, and safety aspects.

10.5 Managing a Test Laboratory

Because of the variety of the tests and of the related test methods, instruments and equipment, the activity of a test center must be organized with the aim of guaranteeing the accuracy, reliability, and repeatability of the measurements.

Fig. 10.5 Data acquisition and transmission systems installed on the tractor cab

The complexity of such an issue will be illustrated by a short description of the organization of the *Centro Prove Macchine Agricole* (CPMA, Testing Centre of Agricultural Machinery) of CRA-ING of Monterotondo (Rome, Italy). CPMA aims at studying the performances of soil tillage machines and their safety characteristics under operative conditions, providing the manufacturers with technical support for developing their products. CPMA also executes the performance and safety tests for the ENAMA (National Body for Agricultural Mechanization) certification and participates in the activity of the ENTAM (European Network for Testing of Agricultural Machinery).

10.5.1 Quality System

CPMA is structured as a technical–scientific laboratory, and has developed a "Quality System" (QS) aimed at obtaining, from the National System for the Crediting of Test Laboratories, the official crediting for the test activity conducted according to the severe requirements of the ISO/IEC 17025:2005 (2005) standard (Pochi et al. 2008). Such a QS requires rationalization of all aspects of the activity.

For each category of machine to be tested it has been necessary to define a "Quality Project" (QP) in which all test procedures (according to the ISO, EN, UNI, and ENAMA standards), the authorized personnel, the required instruments and the conditions for their use, the data processing and the method of presentation of the results are codified and certified. The QP applied in each test is indicated, with the results, in the final report. As described above, a variety of instruments is involved in the tests and the quality of the results depends on their efficiency. According to the QS, the instrument are classified on the basis of their use, as reference instruments (officially certified, class A), control and calibration instruments (class B) and instruments used in the tests (classes C and D). They are stored under controlled conditions of temperature and humidity. Finally, they periodically undergo calibration tests according to the "Operative Instructions", documents in which the calibration activity is strictly planned step by step. For several instruments, it has been necessary to develop proper test benches capable of performing their calibration. The calibration provides an internal certificate that indicates the accuracy and the uncertainty of the instrument and is mentioned in the final test report. These criteria are always adopted in the continuous updating of the instrumental system.

All the activities are performed by qualified personnel properly skilled in the use of the machines involved in the tests (tractors, operating machines) that require correct adjustments. Moreover, they must be capable of correctly installing the sensors on the machines and of operating the calibration tests by means of the related test benches. The QS promotes the realization of courses aimed at updating and improving the skill level of the personnel involved in the activity, both in performance tests and safety matters. The courses represent a period for discussion of the different aspects of the activity, such as the introduction, in the test system, of new instruments and equipment and the individuation of the most critical aspects of their calibration and use in the tests.

10.5.2 Mobile Laboratory

CPMA test activity is based on the use of a "mobile laboratory", an integrated system represented by a *field unit* and a *support unit* managed according the QS (Fanigliulo et al. 2004).

The *field unit* is represented by a tractor equipped with the sensors required by the specific test (among those described in Sect. 10.4.2), a PC with a PCI card handling the acquisition of the data; a LCD monitor, and a photocell system indicating the start and end of measurements on the test plot. As the tractor must be properly coupled to the implement to be tested, in order to satisfy all power requirements three tractors of different power classes have been equipped as described and are used depending on the machine to be tested (Fig. 10.6).

Their efficiency is periodically verified in tests on the dynamometric brake, which also provides updated engine characteristic curves. The curves can also be

Fig. 10.6 Instrumented tractors of three different power classes

used for indirectly determining the total power provided by the engine and the hourly fuel consumption from the average engine speed during the work (Fig. 10.7).

A further element of the field unit is represented by the *traction vehicle* in Fig. 10.3, equipped with a PC similar to those on the tractors and used in traction tests and penetrometric tests.

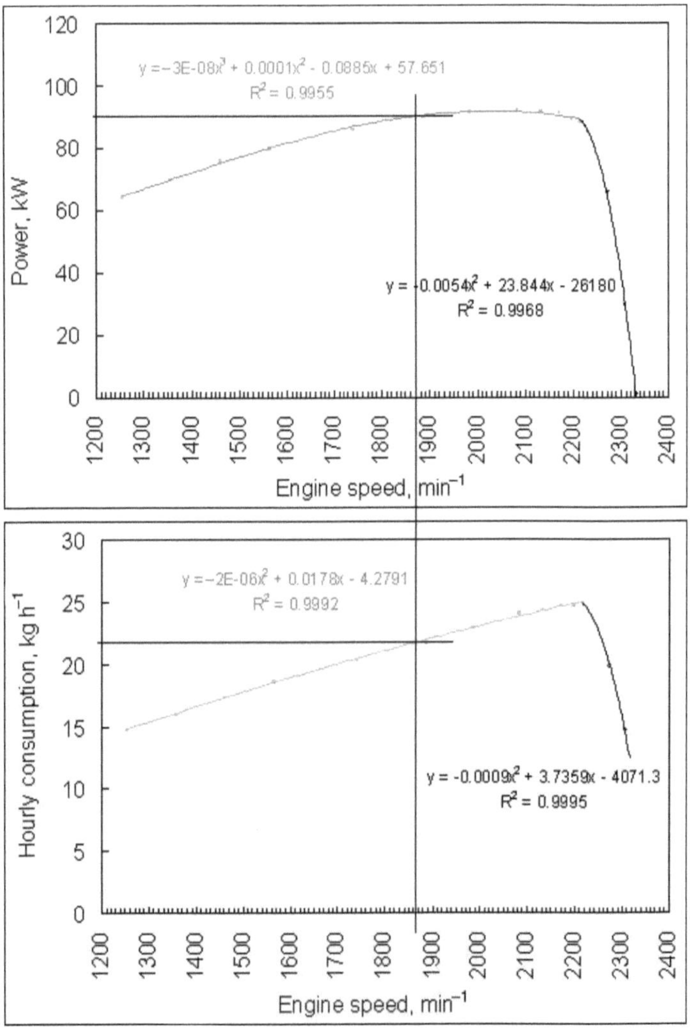

Fig. 10.7 Characteristic curves of power and hourly fuel consumption of one of the tractors of the field unit. The series of data have been divided into homogeneous parts in order to obtain the relative functions of regression. These allow, from an engine speed value, the calculation of the corresponding power and fuel consumption. For this reason, the curves must be continuously updated, taking into account environmental factors (temperature, atmospheric pressure, etc.)

The *support unit* consists of a van equipped as a control station; during the tests it is parked on the border of the field (Fig. 10.8). Its PC exchanges data with the *field unit*'s PC, by means of a radio-modem system, and allows the monitoring, in real time, of the main test parameters, the efficiency of the instruments, etc. The data are immediately processed, providing the test results in a short time. The van also houses the equipment and instruments used for evaluating quality of work, such as the profile-meter, the hand penetrometer, the sieves, etc.

The mobile laboratory is a reliable and effective instrument providing detailed data on the performances of the tested machines. Its limit is represented by the difficulty in performing tests outside of the test center. For instance, because of the particularity of the installation of the torquemeters on the tractors' p.t.o., they can rarely be mounted on tractor models different from those of the field unit; and traction tests always require a traction vehicle (i.e. a second tractor) with suitable dimensions and power. These difficulties can be overcome by equipping the mobile laboratory with sensors of general application, such as the p.t.o. drive shaft instrumented with extensimeter (see Sect. 10.4.2.2) or the frame for direct measurement of the force of traction (see Sect. 10.4.2.1), and increasing the use of miniaturized systems for the radio-transmission of data. An upgrade of the mobile laboratory based on such sensors is being developed.

Fig. 10.8 The support unit and a couple of infra-red barriers used to determine the start and the end of the data acquisition

References

Al-Janobi A (2000) A data-acquisition system to monitor performance of fully mounted implements. J Agric Eng Res 75:167–175

BSI (1975) Methods of test for soils for civil engineering purposes. British Standard Institution, London

Chen Y, Tessier S, Rouffignat J (1998) Soil bulk density estimation for tillage system and soil textures. Trans Am Soc Agric Eng 6:1601–1610

Fanigliulo R, Pochi D, Volpi C, Santoro G (2004) A mobile system to evaluate the performances of agricultural machinery under field conditions. J Agric Eng 4:89–95

Fanigliulo R, Pochi D (2008) In-depth primary soil tillage without inverting: comparison of the operative performances of three subsoilers. Proc Int Conf on Agric Eng, AgEng 2008

Grant CD, Dexter AR, Huang C (1990) Roughness of soil fracture surfaces as a measure of soil microstructure. J Soil Sci 41:95–110

ISO/IEC 17025:2005 (2005) General requirements for the competence of testing and calibration laboratories. Bruxelles, Belgium.

McLaughlin NB, Heslop LC, Buckley DJ, St Amour GR, Compton BA, Jones AM, Van Bogedom (1993) A general purpose tractor instrumentation and data logging system. Trans Am Soc Agric Eng 36(2):265–273

Perdock UD, Kouwenhoven JK (1994) Soil-tool interactions and field performance of implements. Soil Tillage Res 2–4:283–326

Perfect E, McLaughlin NB, Kay BD (1997) Energy requirements for conventional tillage following different crop rotations. Trans Am Soc Agric Eng 40(1):45–49

Peruzzi A, Raffaelli M, Di Ciolo S (1999) Proposal of methodology for the evaluation of the quality of work of the operative machines for soil tillage. J Agric Eng 3:156–167

Pochi D, Ragni L, Santoro G (1996) A mobile laboratory for the analysis of the tractor-implement-soil interactions. Third part: comparison between methodologies to evaluate the power adsorbed by implements. J Agric Eng 27(4):193–197

Pochi D, Cervellini C, Brannetti G, Fanigliulo R (2008) Testing of agricultural machinery at CRA-engineering (Italy) according to the ISO/IEC 17025:2005 standard. Proc Int Conf on Agric Eng, AgEng 2008.

Ragni L, Santoro G (1996) A mobile laboratory for the analysis of the tractor-implement-soil interactions. First part: system description. J Agric Eng 27(2):65–72

Raper RL, Grift TE, Tekeste MZ (2004) A portable tillage profiler for measuring subsoiling disruption. Trans Am Soc Agric Eng 47(1):23–27

Römkens MJM, Wang JY, Darden RW (1988) A laser microreliefmeter. Trans Am Soc Agric Eng 31(2):408–413

Sandri R, Anken T, Hilfiker T, Sartori L, Bollhalder H (1998) Comparison of methods for determining cloddiness in seedbed preparation. Soil Tillage Res 45:75–90

Watts CW, Longstaff DJ (1989) Mobile instrumentation and data processing system for testing field machinery. J Agric Eng Res 43:67–76

Part III
Soil Management and Mechanical Weed Control

Chapter 11
Mechanical Weed Control

Hans W. Griepentrog and Athanasios P. Dedousis

11.1 Introduction

In sustainable production systems the soil fertility is of high importance. The soil and its biological activities are the main components and need to be optimised in relation to seed germination and growth conditions of the intended crop (Lampkin 1994). When discussing tillage aims, the management of soils has always had the secondary aim of controlling weeds.

However, in most cases mechanical weed control operations with soil-engaging tools also increase soil aeration and water infiltration including related positive effects. Mechanical weed control has had increased attention in recent years. The main reason was the intention to reduce herbicide input or completely substitute chemical inputs as required in organic production systems, since consumers demand high quality as well as safe food products and pay special attention to protecting the environment.

Mechanical weeding of row crops has a long tradition. One driver behind the invention of the grain drill was to allow hoeing of grain seedlings between the rows. At that time this operation was new compared with common non-row broadcasted seeding. Today inter-row hoeing is frequently used and the weed control efficacy is highly appreciated and accepted. By using standard hoeing, the remaining challenge is to control weeds within the row (intra-row weeding).

The aim of this chapter is to highlight and focus on novel and promising developments in soil-engaging intra-row hoeing. Through these technological

A.P. Dedousis (✉)
Centre for Research and Technology, Institute of Technology and Management of Agricultural Eco-Systems, Thessaly, Greece
e-mail: dedousis@cereteth.gr

H.W. Griepentrog
University of Copenhagen, Institute of Agriculture and Ecology, Copenhagen, Denmark

A.P. Dedousis and T. Bartzanas (eds.), *Soil Engineering*, Soil Biology 20,
DOI 10.1007/978-3-642-03681-1_11, © Springer-Verlag Berlin Heidelberg 2010

developments such as precise inter- and intra-row weeders, it is possible to control weeds in a way that meets consumer and environmental demands.

11.2 State of the Art

Harrowing and hoeing are the oldest, highly matured and most common non-chemical weeding operations treating the whole surface area and the inter-row area.

Harrowing is mainly used as a whole soil surface treatment, e.g. in grain crops. Machine choice, timing and adjusting is of high importance in order to achieve best results in selectivity to reduce weed density and minimize crop losses (Rasmussen 1992). In row crops, harrowing is currently also used additionally to hoeing to target weeds in the intra-row area (van der Weide et al. 2008). Here again the weeding efficacy is highly dependant on selectivity effects because in most cases the crop plant is treated in the same way as the neighbouring weed plants. In general the weed control efficacy is often unsatisfactory.

Hoeing is at least 120 years old and is still used as a standard weed control operation today. The first hoes were horse-drawnand those today are tractor-mounted or still tractor-pulled. Currently, often a second operator visually controls the hoe laterally by hand. Tines or rotating discs (rotary hoes) are fixed to a frame and penetrate the upper crust of the soil. The treatment is effective on dry, compact soil and a stable working depth is maintained by ground wheels.

Laber (1999) defined and classified hoeing into three control principles:

- Operational principle: Soil-engaged treatment (tillage) between crop rows.
- Physical principle: Soil coverage of weeds, weed root/stem cutting and uprooting of weeds (whole plant or partly).
- Physiological principle: Reduction of photosynthesis and reduction of water transpiration.

As with most mechanical weeding operations, crop plant losses always occur (Vanhala et al. 2004), especially if high weed control efficiencies are aimed at. Crop losses result from soil coverage, crop leaf damage, root damage and disturbance. The standard hoe setting for the untreated crop row strips is 100 mm which gives approximately a maximum of 80% treated area, e.g. in sugar beet. This row band width is measured as a row clearance between the hoe unit tools or shares. Most crop losses occur due to soil disturbance close to crop plants. A conflict of aims appears between (1) maximizing treated area to increase weeding efficiency and (2) minimizing crop losses by keeping a sufficient distance to crop rows. Therefore the adjustment of the hoe unit working width becomes an important factor for achieving an acceptable cultivation result. This result is a compromise between the maximum cultivated area and adequate tolerance when setting the machine in order to avoid crop damage.

Several developments and investigations have been conducted to automate the lateral control of conventional hoes (Tillett 1991; Home 2003). Today the most promising automation principles are based on GPS (Van Zuydam et al. 1995; Dijksterhuis et al. 1998, Griepentrog et al. 2007) and computer vision (Sogaard and Olsen 2003; Aastrand and Baerveldt 2005; Tillett et al. 2008). A fusion of both is seen today as the most promising strategy, because advantages and disadvantages of absolute and relative referencing principles compensate each other (Pilarski et al. 2002; Downey et al. 2003).

Hoeing is used for weed control in row crops mainly in sugar beet, maize and vegetable production. It treats only the soil surface between the crop rows, leaving an untreated band where the crop plants grow within the rows (intra-row area).

New developments in hoeing of the intra-row areas are emerging today (van der Weide et al. 2008). These systems are based on guiding soil-engaged tools into the row when there is no crop plant. This so-called "intelligent hoeing" requires precise positional information on the locations of the crop plants. This information can be provided by cameras mounted on the machine capable of real-time image analysis. These systems can be significantly supported in terms of computational requirements by geo-referenced seed maps retrieved from prior seeding operations (Griepentrog et al. 2005).

By using advanced intra-row hoeing which is non-selective, a third target area arises, the close-to-crop area (Griepentrog et al. 2003). This area has to be untreated by tillage tools because of the high risk of crop plant damage in terms of leaf rupture and root disturbance or even root cuttings (Norremark and Griepentrog 2004). Recent investigations have been conducted by Mathiassen et al. (2006) and Sogaard and Lund (2007) targeting the close-to-crop weeds using laser and micro-spraying.

11.3 Advanced Intra-Row Hoeing

The following text describes intra-row weeders based on computer-controlled soil-engaging tools to treat the intra-row area and avoid crop damage. They all rely on information about the crop plant locations as an input to the control system because they use non-selective tools. Common to all these systems, except the cycloid hoe that uses geo-referenced seeds, is that crop detection is mainly based on image analysis techniques applied on captured pictures of the crop rows taken by cameras.

11.3.1 Blade Inter- and Intra-Row Hoe

A combination of an inter-row and intra-row hoe was developed by Home (2003). It consists of an inter-row "ducks foot" blade that has attached to it reciprocating blades to treat the intra-row weeds (Fig. 11.1). The plants are detected using computer vision which can discriminate between plants and weeds. When the plant is detected that blades are folded in and when there are no plants the motor activates the cam and the blades are folded out. Home (2003) undertook several

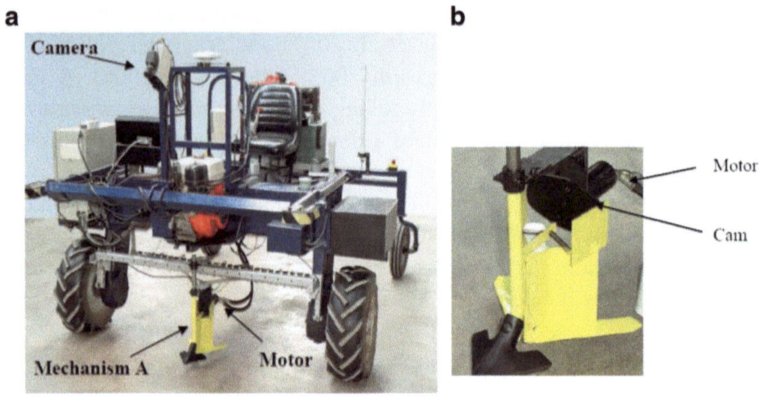

Fig. 11.1 (**a**) Autonomous vehicle with intra-row mechanism (**b**) Blade inter- and intra-row hoe (Home 2003)

field investigations of the proposed system with various intra-row plant spacing and working speeds. He found out that at a spacing of 300 mm the reciprocating intra-row blades avoid entering the root zone (close-to-crop area) up to speeds of 4 km h^{-1}, but at 8 km h^{-1} 17% of the crop root zone was entered. At 250 mm intra-row plant spacing excessive damage occurred with 70% of the crop zone being touched by the intra-row blades, and this was increased when using higher working speeds (Home 2003).

11.3.2 Blade Arm Hoe

An intra-row weed control system which consists of blades mounted on a pivoting arm (Fig. 11.2) has been developed by the company Radis Mechanisation, France. Light sensors sense the within-row crop plants. When no plant is detected the pivoting arm is moved into the intra-row area by an air pressure cylinder, thus cultivating and removing the intra-row weeds. Bakker (2003) reports that at a driving speed of 5 km h^{-1} weeds are removed up to 20 mm from the plant. However Bleeker (2005) reports a maximum speed of 3 km h^{-1} due to the mechanical transition of the intra-row hoe. The system is designed for widely spaced vegetables and the minimum intra-row spacing that the system can work with is 220 mm (Bakker 2003). While testing the weeder, it was limited by the plant detection system. Transplanted crops in which the plant is way ahead of the weeds' growing stage will not present a problem.

Fig. 11.2 The Radis blade arm hoe (Bleeker 2005)

Fig. 11.3 Prototype single row rotating disc hoe (Garford Farm Machinery)

11.3.3 Disc Hoe

The rotating disc-hoe consists of a rotating disc which acts in a horizontal plane and has a cut-out sector and a bevel cut back at its circumference (Fig. 11.3). The disc centre moves at a distance parallel to the crop row so that its swept area passes between the plants and also between the rows.

The bevelled cut-out sector enables the disc to pass the plants without making contact with them and also leaves an undisturbed no-till circular area of soil around them. It disturbs the soil between the no-till areas between the plants, thus providing intra-row weeding. In addition, the disc disturbs and weeds the soil between the rows (Dedousis 2007; O'Dogherty et al. 2007). A vision system detects the phase of approaching plants and that information is combined with measured disc rotation to calculate a phase error between the next plant and disc cut-out. That phase error is corrected by advancing or retarding the hydraulic drive synchronising the mechanism, even in the presence of crop spacing variability (Tillett et al. 2008)

A prototype rotating disc hoe was constructed by Garford Farm Machinery based on a standard vision-guided inter-row steerage hoe. Field investigations on transplanted cabbage planted at an inter- and intra-row spacing of 300 mm have shown this to be a very effective mechanism where approximately 60% of the weeds within an 80 mm radius of the crop were destroyed. At greater radii this increases to 80% of the weed population (Dedousis et al. 2007).

11.3.4 Cycloid Hoe

The University of Applied Sciences Osnabrück, Germany, in collaboration with Amazone Werke, Germany, have developed a mechanical weed control system for intra-row cultivation, principally for maize. A GPS-based electro-hydraulic motor system for the rotor was designed, developed and tested at the University of

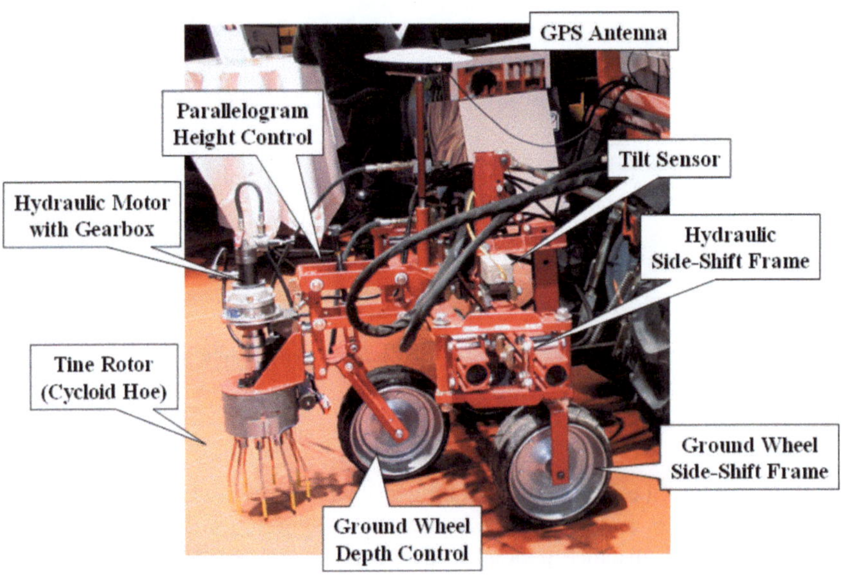

Fig. 11.4 Cycloid hoe (Griepentrog et al. 2006)

Copenhagen (Griepentrog et al. 2006; Norremark et al. 2008). The system is based on geo-referenced seeds derived from the crop planting operation as input data for the real-time controller (Griepentrog et al. 2005).

The individual tines of the rotor can be activated by setting two different tine trajectories in order to move into or stay outside the row. The non-activated tine trajectories can be described as curves traced by a point on the circumference of a circle that rolls on a straight line (cycloid). In general it can be regarded as a superposition of a rotational and translational movement (Fig. 11.4). Main parameters to achieve a particular tillage effect are the ratio of forward speed to rotational speed, the diameter of tine rotation, the lateral offset to crop rows, the number of tines and the shape and design of tine tips. The rotor has eight tines and the tine tips have a rotational diameter of 0.234 m. The weeding effect of the hoe tines is similar to that of hoe shares and is mainly due to uprooting, weed soil coverage and root cutting.

The rotor principle and the rotor itself was developed at the University of Applied Sciences at Osnabrück, Germany (Wißerodt et al. 1999).

Results from field experiments show that the accuracy of tool guidance is high and sufficient for weed cultivations (Norremark et al. 2008). The purely GPS-based system could benefit from adding other sensor types such as optical to be more reliable and accurate.

11.4 Conclusions

Weeding tools for intra-row treatments based on selectivity are limited in their control efficacy. Advanced new tools based on novel crop and weed plant detection and computer control systems show promising results. Further field experiments have to be conducted to improve functionality, reliability and weed control efficacy.

References

Aastrand B, Baerveldt AJ (2005) A vision based row-following system for agricultural field machinery. Mechatronics 15(2):251–269

Bakker T (2003) Autonomous weeding literature review. SCO Report 03–01, Wageningen University: Systems and Control Group.

Bleeker P (2005) First experiments with intra-row weed control with a camera. In: Proceedings of the spatial and dynamic weed measurements and innovative weeding technologies, Bygholm, Denmark, European Weed Research Society

Dedousis AP, Godwin RJ, O'Dogherty MJ, Tillett ND, Grundy AC (2007) Inter and intra-row mechanical weed control with rotating discs. In: Stafford JV (ed) Precision Agriculture 07. Wageningen Academic Publishers, Wageningen, NL, pp 493–498

Dedousis AP (2007) An investigation into the design of precision weeding mechanisms for inter and intra-row weed control. PhD thesis. Cranfield University, Silsoe

Dijksterhuis HL, Van Willigenburg LG, Van Zuydam RP (1998) Centimetre-precision guidance of moving implements in the open field: a simulation based on GPS measurements. Comput Electron Agr 20:185–197

Downey D, Giles D, Slaughter DC (2003): Ground based vision identification for weed mapping using DGPS. In: Proceedings ASAE Annual International Meeting Las Vegas, Nevada, USA, ASAE, paper no. 03-1005

Griepentrog HW, Norremark M, Nielsen H, Blackmore BS (2003) Individual plant care in cropping systems. In: Stafford JV, Werner A (eds) Proceedings 4th European Conference on Precision Agriculture ECPA, 26.6.2003 Berlin. Wageningen Academic Press, Wageningen, NL, pp 247–251

Griepentrog HW, Norremark M, Nielsen J (2006) Autonomous intra-row rotor weeding based on GPS. In: Proceedings CIGR World Congress – Agricultural Engineering for a Better World, 9.2006 Bonn, Germany, VDI, Düsseldorf, Germany

Griepentrog HW, Norremark M, Nielsen J, Soriano Ibarra J (2007) Autonomous inter-row hoeing using GPS based side-shift control. Agricultural Engineering International – The CIGR Ejournal IX(Manuscript ATOE 07 005)

Griepentrog HW, Norremark M, Nielsen H, Blackmore BS (2005) Seed mapping of sugar beet. Precis Agr 6(2):157–165

Home M (2003) An investigation into the design of cultivation systems for inter- and intra-row weed control. EngD Thesis, Cranfield University, Silsoe, UK

Laber, H. (1999): Effizienz mechanischer Unkrautregulationsmaßnahmen im Freilandgemüsebau. Dr. rer. hort. Thesis, University of Hanover

Lampkin N (1994) The living soil. Organic Farming, Chap. 2. Farming Press, Ipswich, UK, pp 13–51

Mathiassen SK, Bak T, Christensen S, Kudsk P (2006) The effect of laser treatment as a weed control method. Biosystems Eng 95(4):497–505

Norremark M, Griepentrog HW (2004) Analysis and definition of the close-to-crop area in relation to robotic weeding. In: Proceedings 6th workshop of the EWRS working group 'Physical and Cultural Weed Control', 8.3.2004 Lillehammer, Norway, p 127–140

Norremark M, Griepentrog HW, Nielsen J, Soegaard HT (2008) The development and assessment of the accuracy of an autonomous GPS-based system for intra-row mechanical weed control in row crops. Biosystems Eng 101(4):396–410

O'Dogherty MJ, Godwin RJ, Dedousis AP, Brighton JL, Tillett ND (2007) A mathematical model of the kinematics of a rotating disc for inter- and intra-row hoeing. Biosystems Eng 96(2):169–179

Pilarski T, Happold M, Pangels H, Ollis M, Fitzpatrick K, Stentz A (2002) The demeter system for automated harvesting. Auton Robots 13:9–20

Rasmussen J (1992) Testing harrows for mechanical control of annual weeds in agricultural crops. Weed Res 32:267–274

Sogaard HT, Lund I (2007) Application accuracy of a machine vision-controlled robotic micro-dosing system. Biosystems Eng 96(3):315–322

Sogaard HT, Olsen HJ (2003) Determination of crop rows by image analysis without segmentation. Comput Electron Agr 38(2):141–158

Tillett ND, Hague T, Grundy AC, Dedousis AP, Tillett ND, Hague T, Grundy AC, Dedousis AP (2008) Mechanical within-row weed control for transplanted crops using computer vision. Biosystems Eng 99(2):171–178

Tillett ND (1991) Automatic guidance sensors for agricultural field machines – a review. J Agric Eng Res 50(3):167–187

Van der Weide RY, Bleeker PO, Achten VTJM, Lotz LAP, Fogelberg F, Melander B (2008) Innovation in mechanical weed control in crop rows. Weed Res 48(3):215–224

Van Zuydam RP, Sonneveld C, Naber H (1995) Weed control in sugar beet by precision guided implements. Crop Prot 14(4):335–340

Vanhala P, Kurstjens D, Ascard J, Bertram A, Cloutier D, Mead R, Raffaelli M, Rasmussen J (2004) Guidelines for physical weed control research – flame weeding, weed harrowing and intra-row cultivation. In: Proceedings 6th EWRS Workshop on Physical and Cultural Weed Control, 8.3.2004 Lillehammer, Norway

Wißerodt E, Grimm J, Kemper M, Kielhorn A, Kleine-Hartlage H, Nardmann M, Naescher J, Trautz D (1999) Gesteuerte Hacke zur Beikrautregulierung innerhalb der Reihe von Pflanzenkulturen. In: Proceedings VDI-Tagung Landtechnik Braunschweig, Germany, VDI-Verlag, Düsseldorf, Germany, pp 155–160

Chapter 12
Characterisation of Soil Variability with Depth Using Vis–NIR Spectra and Chemometric Tools

A.M. Mouazen, R. Karoui, J. De Baerdemaeker, and H. Ramon

12.1 Introduction

Agricultural soils are characterised as non-homogeneous material with substantial variability in the physical and chemical properties. Many studies show large short-range soil variability, not only in large-scale fields (McBratney and Pringle 1997; Corwin et al. 2003; Vrindts et al. 2005; Godwin and Miller 2003), but in small-scale fields too (Mouazen et al. 2003). There are many reasons for this variability within the upper soil layers: namely, topography, soil texture, soil compaction, slope, climate, plant activity and depth of soil from the mother rock. This variability exists not only spatially but within the soil profile (depth) as well. To establish fertilisation recommendations of mineral nitrogen status, the Soil Service of Belgium (Heverlee, Belgium) collects samples from three different layers, d1 (0–30 cm), d2 (30–60 cm) and d3 (60–90 cm), to be subjected to routine analysis for texture and mineral nitrogen. Based on this measurement, they make nitrogen fertilisation recommendations to maintain allowable soil nitrogen levels and reduce groundwater contamination. So far, the traditional laboratory analyses have revealed variability in texture, moisture content and total mineral nitrogen content between the three soil layers. A method for assessing this variability between soil layers using a technique that is fast, cost-effective and accurate may be useful.

A.M. Mouazen (✉)
Natural Resources Department, Cranfield University, Bedfordshire MK43 0AL, UK
e-mail: a.mouazen@cranfield.ac.uk

K. Karoui
Unité de Recherches Typicité des Produits Alimentaires, ENITA Clermont Ferrand, Marmilhat, BP35, 63370, Lempdes , France

J.De Baerdemaeker and H. Ramon
Division of Mechatronics, Biostatistics and Sensors, Faculty of Bioscience Engineering, Kasteelpark Arenberg 30, B-3001, Heverlee , Belgium

A.P. Dedousis and T. Bartzanas (eds.), *Soil Engineering*, Soil Biology 20,
DOI 10.1007/978-3-642-03681-1_12, © Springer-Verlag Berlin Heidelberg 2010

Visible (Vis) and near infrared (NIR) spectroscopy has been proven to be a promising measurement technique for qualitative and quantitative analysis of variability in different agricultural materials (Mouazen et al. 2005a). After excluding the capital cost of the instrument, results obtained by this method are characterised as fast, cheap and sufficiently accurate. Therefore, many researchers in the agricultural and food industry sectors have considered Vis–NIR spectroscopy as an alternative technique to laboratory measurement methods for different products. Examples of quantitating the properties of different agricultural materials using Vis–NIR spectroscopy are numerous (Shenk et al. 1992; Park et al. 1999; Reyns et al. 2001; Paulsen et al. 2003; Paulsen and Singh 2004; De Belie et al. 2003; Saeys et al. 2005; Mouazen et al. 2005b; Reeves and McCarty 2001). Qualitative analyses of agricultural materials with Vis–NIR spectroscopy have also been reported, e.g. classification of cheese (Karoui et al. 2006) and extra virgin olive oil (Downey et al. 2003) to different geographical origins, honey (Corbella and Cozzolino 2005) according to their flora origin, and separation of wheat semolina samples into three levels of water concentration (Devaux et al. 1988). Most of these studies used the combination of Principal Component Analysis (PCA) and Factorial Discriminant Analysis (FDA) to develop their qualitative models. However, very limited numbers of similar studies on qualitative analysis of soils have been reported. This can be attributed to the high variability existing in soils which makes classification rather a difficult task, particularly when dealing with large geographic areas and different soil depths. Therefore, only PCA has been used to classify soil spectra into different groups (Stenbergh et al. 1995; Leone and Sommer 2000; Chang et al. 2001; Boer et al. 1996). However, PCA is a descriptive and not a predictive technique that can classify new individuals into *prior* established groups. The feasibility of using predictive FDA together with PCA for discriminating between different soils with large variability has not so far been investigated.

The scope of this paper is the use, in soil, of FDA for classifying Vis–NIR spectra of soil samples taken at three different depths from fields covering large areas in Belgium and northern France.

12.2 Materials and Methods

12.2.1 Soil Samples

Soil samples were collected from multiple fields distributed in Belgium and the northern part of France. A total of 237 samples was obtained from the Soil Service of Belgium (Heverlee, Belgium). They were collected in the late winter and spring of 2004 from 112 fields at three different depths of 0–30 cm, 30–60 cm and 60–90 cm. These fields represent a wide range of variable soil textures, colours, origins and other physical and chemical properties. Collecting the soil samples during the wet season resulted in samples with a small range of natural variation in

Table 12.1 Sample statistics for moisture content and total mineral nitrogen of the three-depth soil samples

Property	Number of samples	Minimum	Maximum	Mean	Variance	Standard deviation
d1[a] (0–30 cm)						
Moisture content ($kg\ kg^{-1}$)	112	0.129	0.281	0.191	0.0008	0.028
Total mineral nitrogen ($kg\ ha^{-1}$)	112	1.350	24.440	6.684	17.687	4.206
d2 (30–60 cm)						
Moisture content ($kg\ kg^{-1}$)	103	0.083	0.298	0.195	0.0009	0.030
Total mineral nitrogen ($kg\ ha^{-1}$)	103	2.470	80.830	13.786	165.984	12.883
d3 (60–90 cm)						
Moisture content ($kg\ kg^{-1}$)	22	0.123	0.234	0.175	0.0006	0.025
Total mineral nitrogen ($kg\ ha^{-1}$)	22	2.820	16.370	6.781	13.730	3.705

[a]d is depth

moisture content (Table 12.1). Soil samples were stored in plastic bags at 4°C until laboratory reference and Vis–NIR spectroscopic measurement.

12.2.2 Reference Measurement

Laboratory analyses of mineral nitrogen, texture and moisture content were performed by the Soil Service of Belgium (Heverlee, Belgium), using the official analysis methods (Vanden Auweele et al. 2000). An extremely long time is needed for the determination of soil texture by a combination of wet sieve and hydrometer tests. Therefore, texture was determined in a sensory way by a soil surveyor performing the test with fingers and thumb (White 1997). The soil samples were classified into different texture classes according to the Belgium classification of soil texture, as given in Table 12.2 (Deckers 1996). The gravimetric soil moisture content expressed in $kg\ kg^{-1}$ was measured after oven drying at 105°C for 24 h. The wet soil weight was directly obtained after sample scanning with the spectrophotometer. The overall sample statistics for the investigated data set are given in Tables 12.1 and 12.2.

12.2.3 Spectrophotometer and Scanning

A mobile, fibre-type Vis and NIR spectrophotometer developed by the Zeiss Company (Zeiss Corona 45 visnir fibre) was used. It is fast, precise and robust, without moving parts, which made it suitable for use on vehicles to measure water

Table 12.2 Soil texture at the three soil layers, determined according the Belgium classification of soil texture[28]

Property	Textures	Number of samples
d1[a] (0–30 cm)	Coarse sand	4
	Fine sand	1
	Loamy sand	6
	Light sandy loam	7
	Sandy loam	18
	Light loam	40
	Loam	34
	Clay	2
d2 (30–60 cm)	Loamy sand	3
	Light sandy loam	1
	Sandy loam	4
	Light loam	36
	Loam	48
	Heavy clay	11
d3 (60–90 cm)	Fine sand	1
	Light sandy loam	3
	Light loam	11
	Loam	5
	Clay	2

[a]d is depth

content in the field (Mouazen et al. 2005b). In addition to the InGaAs diode-array for measurement in the NIR region (944.5–1710.9 nm), a Si-array is available for the measurement in the visible and short infrared wavelength region (306.5–1135.5 nm). The light source is a 20 W tungsten halogen lamp illuminating the targeted soil surface. The light illumination and reflectance fibres were collected together at a 45° angle.

Different amounts of fresh soil according to different textures were packed in a plastic cup 1.0 cm deep and 3.6 cm in diameter. No pre-treatment was considered, except removing the plant roots. Soil in the cup was first shaken and a gentle pressure was applied to the surface before the surface was carefully levelled in order to obtain a maximum amount of reflected light. Three reflectance readings were taken from each soil specimen at three different spots by rotating the plastic cups through 120°. Each spectrum was an average of five successive spectra measured at 2.5 s. An average spectrum, representing 15 spectra, was then obtained from the three measured spectra to be considered for spectra processing and statistical analysis.

12.2.4 Principal Component Analysis (PCA)

The PCA was applied on the Vis–NIR spectra recorded on soil samples in order to extract information on three different depth groups. The PCA transforms the original independent variables (wavelengths) into new axes, or principal components (PCs). These PCs are orthogonal, so that the data sets presented on these axes are uncorrelated with each other (Jolliffe 1986; Martens and Naes 1989). Therefore, the PCA expresses the total variation in the data set in only a few PCs and each successively derived PC expresses decreasing amounts of the variance. The first PC covers as much of the variation in the data as possible. The second PC is orthogonal to the first PC and covers as much of the remaining variation as possible, and so on. By plotting the PCs, one can view interrelationships between different variables, and detect and interpret sample patterns, groupings, similarities or differences. The spectral patterns corresponding to the PCs provide information about the characteristic peaks, which are the most discriminating for the samples observed on the map. While similarity maps allow comparison of the spectra in such a way that two neighbouring points represent two similar spectra, the spectral patterns exhibit the absorption bands that explain the similarities observed on the maps.

Before carrying out the PCA, the Multiple Scatter Correction (MSC) was applied to soil Vis–NIR spectra. Then, the PCA was applied to the Vis–NIR spectra to investigate differences between soil depth groups. A total of 148 wavelength points (variables) were included in the PCA. The PCA was carried out using StatBoxPro software (Grimmer Logiciels, Paris, France).

12.2.5 Factorial Discriminant Analysis (FDA)

The aim of the FDA analytical technique is to predict the membership of an individual soil sample following the definition of three qualitative groups created according to the three soil depths. The FDA is a multivariate method that allows the testing of hypotheses (Le bart et al. 1977). It belongs to the field of decisional statistics, and is based on the comparison of multidimensional intra-group variances with inter-group variance. These methods can show the presence of certain relationships between a qualitatively explained criterion and a group of quantitative explanatory characters, and they allow one to describe these latter relationships. The introduction of a qualitative variable within a population allows the division of this population into different groups, with each individual assigned to one group. Discrimination of the groups consists of maximising the variance between their centres of gravity; one can then clarify the properties that distinguish the different groups. If the individual is close to the centre of gravity of its group, it is "correctly classified". If the distance to the centre of gravity of its group is greater than that to the centre of gravity of another group, the individual is "poorly classified" and it will be reassigned to this other group.

The FDA was performed on the first five PCs resulting from the PCA applied to Vis–NIR spectral data recorded on soil samples. Considering the first five PCs only was justified by the fact that they cover the most variation contained in the raw data. The Vis–NIR spectral collections were divided into two data sets for calibration and validation. Two thirds of the samples were used for the calibration set and one third for the validation set. The FDA was carried out using StatBoxPro software (Grimmer Logiciels, Paris, France).

Finally, the first five PCs of the PCA performed on Vis–NIR spectra recorded on soils and the two PCs of the PCA performed on texture, moisture and total mineral nitrogen contents were pooled (concatenated) into one matrix, and this new table was analysed by the FDA. The process consists of gathering the 5 PCs of the PCA performed on the Vis–NIR and the two PCs of the PCA performed on physical parameters into one matrix to take into account the total information collected from the soils. This concatenation approach helps to improve the discrimination between soil samples according to depth by using the Vis–NIR spectra and physical parameters, as well as to assess the ability of this new technique (concatenation) to determine the nature of soils.

12.3 Results and Discussion

12.3.1 *Normalised Spectra*

The averaged, MS corrected spectra of soil samples at three depths are shown in Fig. 12.1, revealing differences in the shape of spectra according to the depth of sampling. The deep and wide trench of water absorption due to the O–H first overtone in the second overtone region at 1450 nm of the three spectra indicates

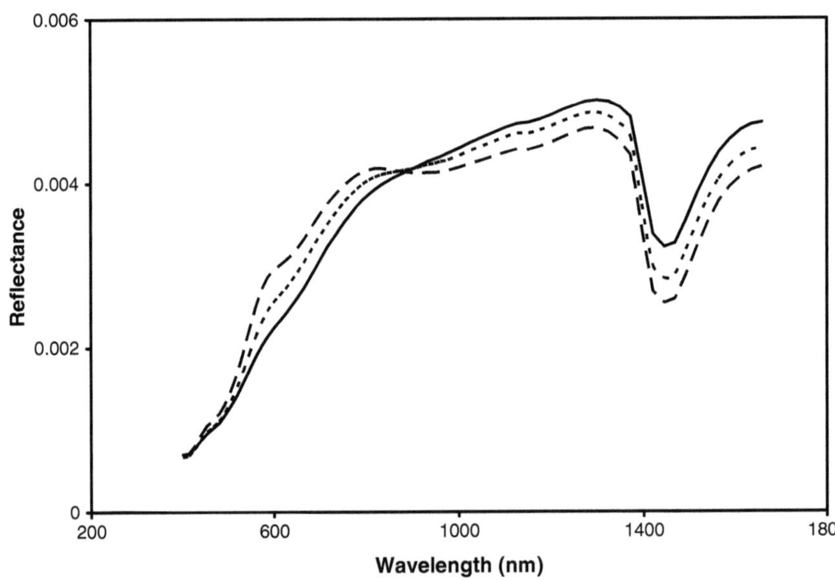

Fig. 12.1 Visible (Vis) and near infrared (NIR) spectra after Multiple Scatter Correction (MSC) recorded between three soil sample groups representing three different depths of d1 (0–30 cm) (*solid lines*), d2 (30–60 cm) (*dotted lines*) and d3 (60–90 cm) (*dashed lines*)

that soil samples were wet in general during the time of scanning. This is in line with the sample statistics for moisture content given in Table 12.1. This wet soil condition at the three depths is attributed to the time of collection of the samples during the late winter and spring of 2004, with large amounts of water incorporated into the soil. However, the deepest layer d3 of 60–90 cm is rather less wet that the upper first layer d1 of 30–60 cm and second layer d2 of 60–90 cm (Table 12.1). Since the difference in moisture content between the three soil depths is rather small (Table 12.1), the variability found in the shape of spectra due to different moisture content is expected to be relatively small. However, the difference between the three averaged spectra is not attributed only to moisture content but also to other properties such as colour, texture and chemical properties. Examining the Vis region (400–750 nm) shown in Fig. 12.1 indicates that the spectrum of the upper layer (d1) has the lowest reflectance (highest absorption). This reflectance increases with increasing depth of sampling, which can be attributed to the darkness of the soil decreasing with depth. The upper layer is rich in bioactivity (plant and animals), and hence it is expected to be also rich in organic matter content, adding darkness to the soil colour. Two peaks of reflectance appear at 600 and at 800 nm, which result from the absorption in the blue region around 450 nm, in the red region around 680 nm and the water absorption in the third overtone region (950 nm). However, these two peaks are distinguishable only when the red colour is encountered. It is obvious that these two peaks are more pronounced for d3 (60–90 cm) compared to the other two layers. The reason for that is the presence of clear yellowish and reddish colours in the deep soil that can be attributed to iron oxides. These yellowish and reddish colours of the deepest soil samples were very visible to the examiner during sample preparation for the optical measurement. Due to the dark colour, the absence of the red colour and the high absorption around the blue band of the upper soil samples (Fig. 12.1), these two peaks can hardly be distinguished for d1. The effect of soil colour on different shapes of spectra of the three groups of sample seems to be large enough to establish shape differences with depth (Fig. 12.3).

12.3.2 Optical Separation Between Different Depth Groups

In order to take into account the total information contained in the Vis–NIR spectral data, PCA was applied to soil spectra. The similarity map defined by PC1 and PC4 accounted for 86.1% of the total variance, with PC1 accounting for 84.9% of the total variance (Fig. 12.2a). This figure shows overlapping between the three groups of samples, particularly at the borders. However a trend towards a good separation between d1 and d3 soils was observed according to PC1.

The spectral patterns associated with the PCs provide the characteristic wavelengths that may be used to discriminate between spectra. Both spectral patterns 1 and 4, associated with PC1 and PC4, respectively, show significant wavelength at 600 nm in the positive range, which is associated with the reflectance peak between the blue (450 nm) and red (680 nm) colours (Figs. 12.1 and 12.2b). The PC1 shows two

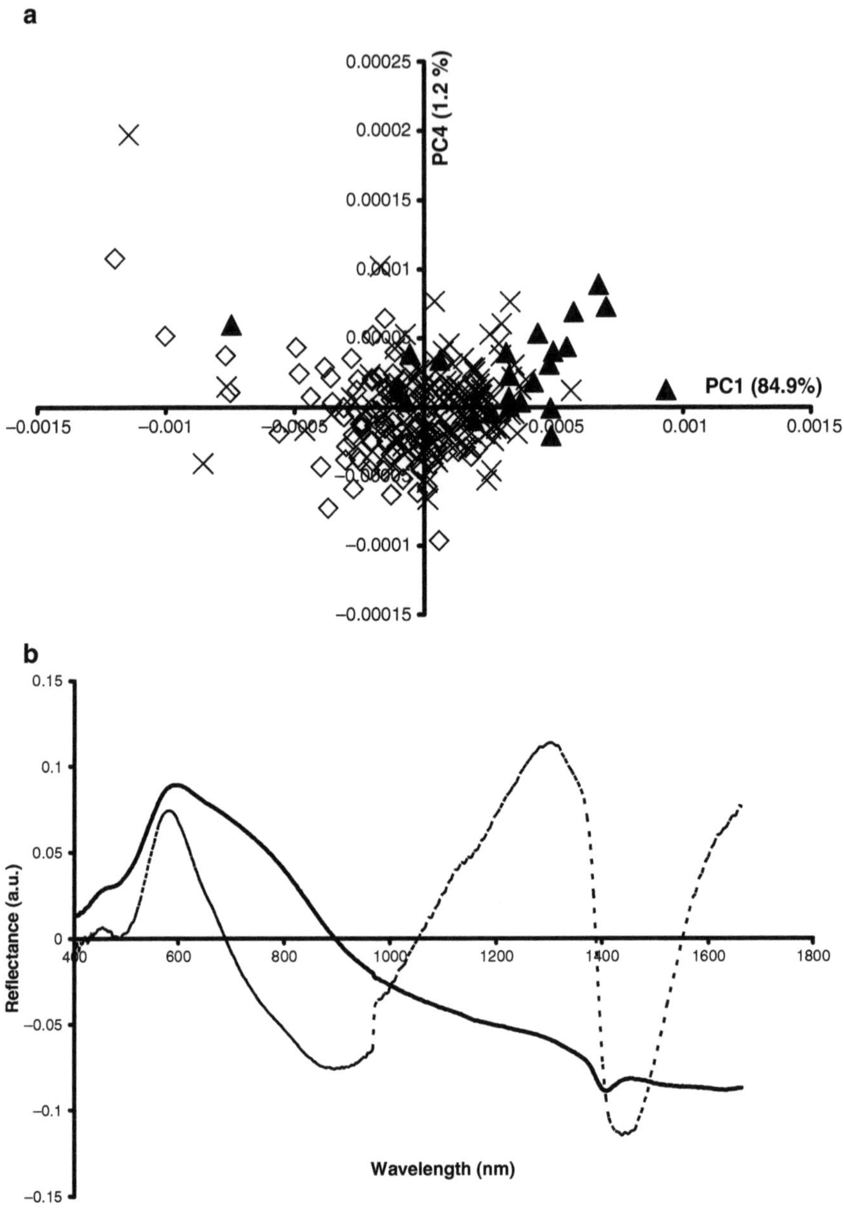

Fig. 12.2 (**a**) Principal Component Analysis (PCA) similarity map determined by principal components one (PC1) and four (PC4) and (**b**) spectral patterns one (*solid lines*) and four (*dashed lines*) for the visible-near infrared (Vis–NIR) spectra of d1 (0–30 cm) (*open diamond*), d2 (30–60 cm) (*times*) and d3 (60–90 cm) (*filled triangle*) soils

negative significant wavelengths at 950 and 1450 nm, those associated with water absorption bands in the third and second overtone regions, respectively. A fourth positive significant wavelength is shown at about 1310 nm, which is associated with the reflectance peak between the two peaks of water absorption (950 and 1450 nm).

In a second step, the ability of Vis–NIR spectra to differentiate soil samples according to different depths was investigated by applying FDA to the first five PCs of the PCA performed on Vis–NIR spectra. Before applying FDA, the three soil groups (d1, d2 and d3) were defined for the investigated soils. The map defined by the discriminant factors 1 and 2 took into account 100% of the total variance, with discriminant factor 1 accounting for 93.4%, as shown in Fig. 12.3. Considering the discriminant factor 1, d1 samples had almost positive score values while d3 soils had negative score values. The d2 samples spread over the positive and negative side of F1. However, there is still visible overlapping between the three groups, particularly between d1 and d2.

Tables 12.3 and 12.4 provide the classification results for the calibration and validation data sets, respectively. Overall correct classification (CC) of 80.2 and 73.4% was observed for the calibration and validation samples, respectively. Among the three different groups, the best classification was obtained for the upper sample set (d1) for both the calibration and validation data sets. This is expected as d1 is rich in organic matter content (plant and animal) compared to the

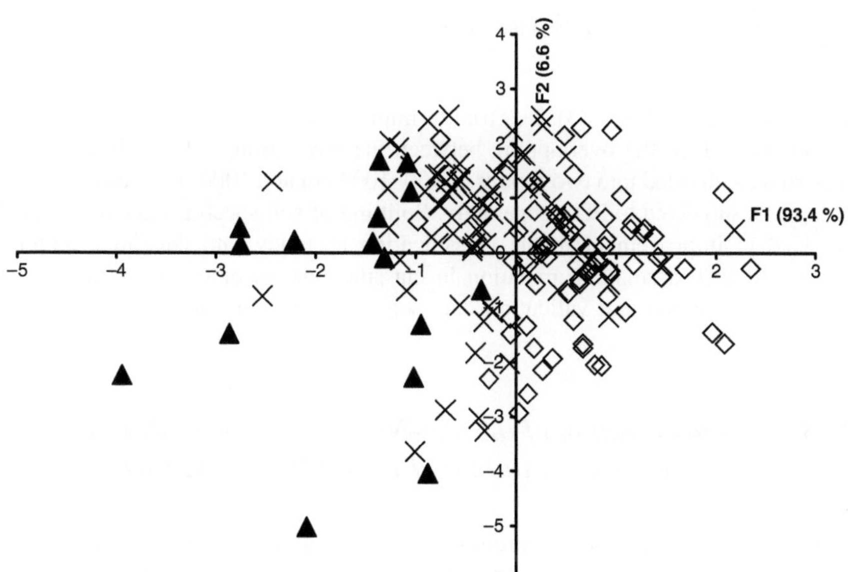

Fig. 12.3 Discriminant analysis similarity map for the calibration data set determined by discriminant factors 1 (F1) and 2 (F2) for the Factorial Discriminant Analysis (FDA) performed on the first five principal components (PCs) of the PCA performed on the visible-near infrared (Vis–NIR) spectra of d1 (0–30 cm) (*open diamond*), d2 (30–60 cm) (*times*) and d3 (60–90 cm) (*filled triangle*) soils

Table 12.3 Classification table for the visible (Vis) and near infrared (NIR) spectra of the calibration data set for three depths of d1 (30–60 cm), d2 (60–90 cm) and d3 (90–120 cm)

Real[a]	Predicted[b]			
	d1	d2	d3	% well classified
Entire spectral region				
d1[c]	66	9	–	88
d2	12	52	7	73.2
d3	–	4	12	75
Total	–	–	–	80.2
400–1000 nm region				
d1	58	17	–	77.3
d2	10	57	4	80.3
d3	1	5	10	62.5
Total	–	–	–	77.2
1000–1650 nm region				
d1	64	9	2	85.3
d2	11	51	9	71.8
d3	1	4	11	86.8
Total	–	–	–	77.8
Concatenation				
d1	70	4	1	93.3
d2	11	52	6	75.4
d3	1	1	13	86.7
Total	–	–	–	84.9

[a]The number of real soils
[b]The number of soils predicted from the model
[c]d is depth

other two layers (d2 and d3). In a trial to improve the accuracy of spectra classification and reduce the overlapping between the three groups shown in Fig. 12.3, spectra were divided into two groups of 400–1000 nm and 1000–1650 nm, and each group was subjected to FDA separately. Splitting of soil spectra into two parts did not lead to improvement in the classification accuracy into the three assigned groups. In fact, a small degradation in classification accuracy was obtained for both the calibration and validation data sets, as shown in Tables 12.3 and 12.4, respectively.

12.3.3 Concatenation of the Vis–NIR Spectra with Moisture Content, Soil Texture and Total Mineral Nitrogen

The concatenation aimed at improving the classification accuracy of soil samples into the three assigned depth groups. The method was applied to the first five PCs of the PCA performed on Vis–NIR spectra of soil and the first two PCs of the PCA performed on soil texture, moisture content and total mineral nitrogen. The seven PCs were gathered into one matrix, which was reanalysed by FDA.

The map defined by the discriminant factors 1 and 2 accounts for 100% of the total variance with discriminant factor 1 accounting for 79.1% (Fig. 12.4a). Considering

Table 12.4 Classification table for the visible (Vis) and near infrared (NIR) spectra of the validation data set for three depths of d1 (30–60 cm), d2 (60–90 cm) and d3 (90–120 cm)

Real[a]	Predicted[b]			
	d1	d2	d3	% well classified
Entire spectral region				
d1[c]	34	3	–	91.9
d2	7	20	8	57.1
d3	1	2	4	57.1
Total	–	–	–	73.4
400–1000 nm region				
d1	24	13	–	64.9
d2	6	24	5	68.6
d3	–	4	3	42.9
Total	–	–	–	64.6
1000–1650 nm region				
d1	35	1	1	94.6
d2	16	14	5	40
d3	1	3	3	42.9
Total	–	–	–	65.8
Concatenation				
d1	33	2	2	89.2
d2	6	20	8	58.8
d3	1	1	5	71.4
Total	–	–	–	74.4

[a]The number of real soils
[b]The number of soils predicted from the model
[c]d is depth

the discriminant factor 1, the upper soil samples d1 were observed mostly on the positive side, whereas the other two soil groups were located mostly on the negative side. This was valid for both the calibration (Fig. 12.4a) and validation (Fig. 12.4b) data sets. Tables 12.3 and 12.4 show slight improvement of the classification accuracy for the calibration and validation spectra, respectively. The overall CC of 84.9 and 74.4% was obtained for the calibration and validation samples, respectively. However, this improvement was evaluated as insufficient for the d2 soil samples, since only 58.8% CC was obtained for the validation data set (Table 12.4). Although the statistical technique of concatenation did not allow for valuable amounts of CC for all groups, the methodology consisting of coupling of Vis–NIR together with soil moisture content, texture and total mineral nitrogen is relevant to discriminate soils collected from three different depths over a large geographic area.

12.4 Conclusions

Although agricultural soils are extremely non-homogeneous materials, those sources of variability related to depth can be extracted using Vis and NIR spectroscopy coupled with chemometric tools. It was shown that the combination of PCA

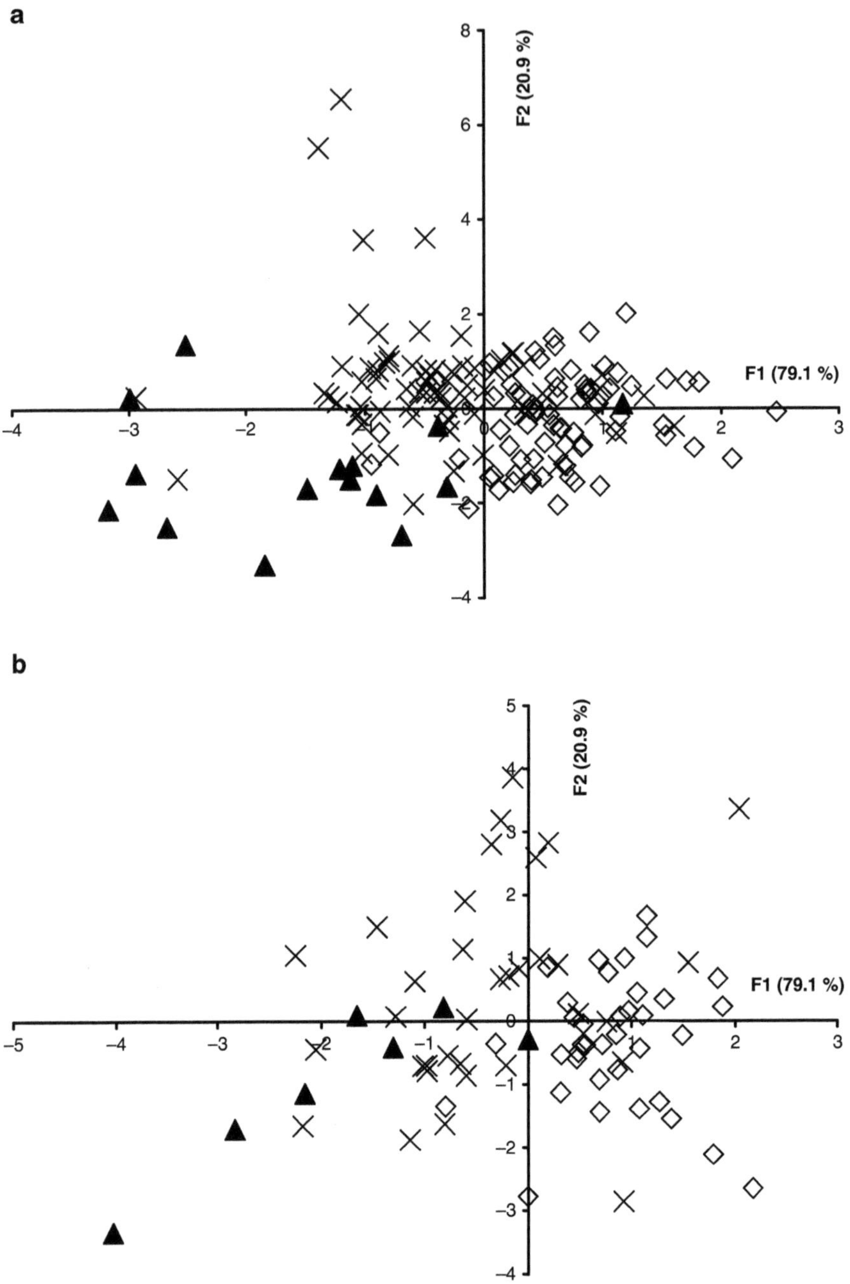

Fig. 12.4 Discriminant analysis similarity map for the calibration (**a**) and validation (**b**) data set determined by discriminant factors 1 (F1) and 2 (F2). Factorial Discriminant Analysis (FDA) performed on the first five concatenated PCs corresponding to the PCA performed on the visible-near infrared (Vis–NIR) spectra, and the two PCs of the PCs performed on texture, moisture content and total mineral nitrogen of d1 (0–30 cm) (*open diamond*), d2 (30–60 cm) (*times*) and d3 (60–90 cm) (*filled triangle*) soils

and FDA is a useful statistical technique that can separate soil samples into three different depth groups of 0–30, 30–60 and 60–90 cm. Dividing the entire spectra into two separate spectral regions of 400–100 nm and 1000–1650 nm did not improve the accuracy of classification of soil samples into the three investigated groups. Utilising information on moisture content, soil texture and total mineral nitrogen by using a concatenation technique led to moderate improvement of the classification accuracy. Information on soil colour and organic matter might be helpful to provide further improvement of the results, as these two properties that vary with depth affect the shape of spectra. This suggests that Vis–NIR spectroscopy coupled with the chemometric techniques described in this study can be considered as a potential technique to classify soils into different classes of colour, moisture, texture, origins and degree of erosion, which might very useful to replace the current methods which are laboratory-based, expensive, costly in time and require operator experience.

Acknowledgment The authors gratefully acknowledge Mr. Jan Bries and Mr. Danny Van de Put at the Soil Service of Belgium for analysing and sharing the soil samples.

References

Boer M, Del Barrio G, Puigdefábregas J (1996) Mapping soil depth classes in dry Mediterranean areas using terrain attributes derived from a digital elevation model. Geoderma 72:99–118

Chang C-W, Laird DA, Mausbach MJ, Hurburgh CR (2001) Near-infrared reflectance spectroscopy–principal components regression analyses of soil properties. Soil Sci Soc Am J 65:480–490

Corbella E, Cozzolino D (2005) Short Communication: the use of visible and near infrared spectroscopy to classify the floral origin of honey samples produced in Uruguay. J Near Infrared Spectrosc 13:63–68

Corwin DL, Kaffka SR, Hopmans JW, Mori Y, van Groenigen JW, van Kessel C, Lesch SM, Oster JD (2003) Assessment and field-scale mapping of soil quality properties of a saline-sodic soil. Geoderma 114:231–259

De Belie N, Pedersen DK, Martens M, Bro R, Munck L, De Baerdemaeker J (2003) The use of visible and near-infrared reflectance measurements to assess sensory changes in carrot texture and sweetness during heat treatment. Biosyst Eng 85:213–225

Deckers J (1996) Bodemgeografie. [Soil Geography]. Katholieke Universiteit Leuven, Leuven, Belgium

Devaux MF, Bertrand D, Robert P, Qannari M (1988) Application of multidimensional analyses to the extraction of discriminant spectral patterns from NIR spectra. Appl Spectrosc 42:1015–1119

Downey G, McIntyre P, Davies AN (2003) Geographic classification of extra virgin olive oils from the Eastern Mediterranean by chemometric analysis of visible and near-infrared spectroscopic data. Appl Spectrosc 57:158–163

Godwin RJ, Miller PCH (2003) A review of the technologies for mapping within-field variability. Biosyst Eng 84:393–407

Jolliffe IT (1986) Principal component analysis. Springer, New York

Karoui R, Mouazen AM, Dufour É, Pillonel L, Schaller E, De Baerdemaeker J, Bosset J-O (2006) Chemical determination of European Emmental cheeses by near infrared spectroscopy using chemometric tools. Int Dairy J 16:1211–1217

Le bart L, Morineau A, Tabard N (1977) Technique de la Description Statistiques Methodes et Logiciel pour l'analyse des . Grand Tableaux, Dunod, Paris

Leone AP, Sommer S (2000) Multivariate analysis of laboratory spectra for the assessment of soil development and soil degradation in the southern Apennines (Italy). Remote Sens Environ 72:346–359

Martens H, Naes T (1989) Multivariate calibration. Wiley, Chichester

McBratney A, Pringle MJ (1997) Spatial variability in soil – implications for precision agriculture. In: Stafford JV (ed) Precision agriculture '97. Bios Scientific Publishers, Oxford, pp 3–31

Mouazen AM, Dumont K, Maertens K, Ramon H (2003) Two-dimensional prediction of spatial variation in topsoil compaction of a sandy loam field-based on measured horizontal force of compaction sensor. Soil Till Res 74:91–102

Mouazen AM, De Baerdemaeker J, Ramon H (2005a) Towards development of on-line soil moisture content sensor using a fibre-type NIR spectrophotometer. Soil Till Res 80:171–183

Mouazen AM, Saeys W, Xing J, De Baerdemaeker J, Ramon H (2005b) Near infrared spectroscopy for agricultural materials: an instrument comparison. J Near Infrared Spectrosc 13:87–98

Paulsen MR, Pordesimo LO, Singh M, Mbuvi SW, Ye B (2003) Maize starch yield calibrations with near infrared reflectance. Biosyst Eng 85:455–460

Paulsen MR, Singh M (2004) Calibration of a near-infrared transmission grain analyser for extractable starch in maize. Biosyst Eng 89:79–83

Park RS, Agnew RE, Arnes RJ (1999) The development of near infrared reflectance spectroscopy calibrations for undried grass silage and their transfer to another instrument using multiple and single sample standardisation J. Near Infrared Spectrosc 7:117–131

Reeves JB, McCarty JW (2001) The potential of near infrared reflectance spectroscopy as a tool for spatial mapping of soil composition for use in precision agriculture. J Near Infrared Spectrosc 9:25–34

Reyns P, Maertens K, De Baerdemaeker J (2001) On-line measurement of corn moisture content on a combine harvester using NIR technology. In: Grenier G, Blackmore S (eds) The 3rd European conference on precision agriculture. Agro Montpellier, Montpellier, France, pp 851–856

Saeys W, Mouazen AM, Ramon H (2005) Potential for onsite and online analysis of pig manure using visible and near infrared reflectance spectroscopy. Biosyst Eng 91:393–402

Shenk JS, Workman JJ Jr, Westerhaus MO (1992) Applications. In: Burns DA, Ciurczak LW (eds) Handbook of near-infrared analysis. Marcel Dekker, New York, pp 419–474

Stenbergh B, Nordkvist E, Salomonsson L (1995) Use of near infrared reflectance spectra of soils for objective selection of samples. Soil Sci 159:109–114

Vanden Auweele W, Boon W, Bries J, Coppens G, Deckers J, Elsen F, Mertens J, Vandendriessche H, Ver Elst P, Vogels N (2000) De chemische bodemvruchtbaarheid van het Belgishe akkerbouw- en weilandareaal. [The chemistry of soil fertility of Belgium Arable and grass lands]. Belgium Soil Service Department, Heverlee, Belgium

Vrindts E, Mouazen AM, Reyniers M, Martens K, Maleki MR, Ramon H, De Baerdemaeker J (2005) Management zones based on correlation between soil compaction, yield and crop data. Biosyst Eng 92:419–428

White RE (1997) Principles and practice of soil science, the soil as a nature resource. Blackwell, Oxford

Chapter 13
Considerations for Recycling of Compost and Biosolids in Agricultural Soil

Maria I. Kokkora, Diogenes L. Antille, and Sean F. Tyrrel

13.1 Introduction

The recycling of organic waste back to the soil is of paramount importance. The utilisation of organic waste within agriculture is encouraged due to the role of organic matter in the soil. Marmo (2008) underlined the importance of soil organic matter (SOM) in relation to climate change. Carbon is a major component of SOM which plays a major role in the global carbon cycle; decreasing the SOM is associated with increased emissions of carbon dioxide (CO_2) and decrease in the soil organic carbon pool. Also, from an agricultural point of view, decrease of the SOM results in degradation of soil fertility, mainly associated with a reduction in the water retention capacity of the soil with increased drought, flooding and erosion risks (Van-Camp et al. 2004), and the capacity of soil to sustain plant roots (Marmo 2008).

EU regulations promote the re-use and recycling of organic materials and attempt to treat them as a resource rather than a waste (Martinez et al. 2007). Following the ban imposed on sea disposal of sewage sludge in 1998 and the EU Landfill Directive 1999/31/EC, the recycling of sewage sludge and biodegradable municipal solid waste (BMSW) became priority for the public and governments. Although incineration is also important, it is expensive and requires continuous investment in incineration capacity to deal with increasing production of both biowastes and sewage sludge (Le 2005).

BMSW (referring to food, garden, paper and cardboard waste produced from households, as well as other waste of a similar nature, e.g. catering waste) and sewage sludge need to be treated prior to application to agricultural land in order to

M.I. Kokkora (✉)
Technological Research Center of Thessalia, T.E.I. of Larissa, 41110, Larissa , Greece
e-mail: mkokkora@trc-thessalia.gr

D.L. Antille and S.F. Tyrrel
School of Applied Sciences, Cranfield University, Bedfordshire MK43 0AL, UK

A.P. Dedousis and T. Bartzanas (eds.), *Soil Engineering*, Soil Biology 20,
DOI 10.1007/978-3-642-03681-1_13, © Springer-Verlag Berlin Heidelberg 2010

stabilise them and minimise pathogens. In 2004–2005, the amount of compost produced in the UK was approximately 1.4 million tonnes compared to only 0.8 million tonnes in 2000–2001. This production of compost comes mainly from the green waste component of BMSW. Approximately 26% of this compost was used in agriculture and 31% in horticulture (Boulos et al. 2006). In addition, the UK produces approximately 1.5 million tonnes (dry solid basis) of sewage sludge (biosolids) per annum of which 60% is treated to standards suitable for agricultural recycling (DEFRA 2007).

Agricultural recycling of compost and biosolids is often regarded as the best practicable environmental option (EA 2004; EC 2007). However, one of the main concerns regarding the use of compost and biosolids in agriculture is with respect to their fertiliser value. Although a vast number of works (Willett et al. 1986; He et al. 1992; Hartl and Erhart 2005; Gibbs et al. 2007 among others) have recognised the benefits of applying compost and biosolids to agricultural land, their use has often been restricted by their variable chemical composition (Sommers et al. 1976; Zmora-Nahum et al. 2007) and the fact that the amount of nitrogen available following application has not been clearly determined (Bowden and Hann 1997; Amlinger et al. 2003). Specifically for biosolids, the nutrient composition can prove to be different depending upon the nature of the wastewater, the wastewater treatment, and the type of processes used during the treatment (EC 2003; Mugnier et al. 2001). The composition of compost varies due to seasonal and geographical variations in the feedstock sources and proportions used, differences in climatic conditions, composting methods and degree of compost maturity (Eriksen et al. 1999; Hargreaves et al. 2008). In field applications, heavy metals and nutrient content of compost and biosolids need to be taken into account as regards fertiliser, soil protection and water laws in each country (Hogg et al. 2002).

In the UK, the maximum allowable quantities of heavy metals in compost and biosolids are provided in the BSI PAS 100 standards (BSI 2005) and the Sludge (Use in Agriculture) Regulations (CEC 1986; CIWEM 1995), respectively. The land application of organic materials needs to abide by the rules set within the codes of good agricultural practice to protect soil, water, and air. As a general rule, the application of organic materials to land is limited to a maximum of 250 kg ha^{-1} of total nitrogen per year (MAFF 1998). However, the timing and the quantity of nitrogen that becomes available to the crop is difficult to predict. Inappropriate application rates can result either in decreased crop production (Svensson et al. 2004) or increased risk of nitrate leaching (Mamo et al. 1999). Furthermore, recommendations of application rates are usually given on the basis of plant available nitrogen. This, in turn, may lead to build-up of soil phosphorus in the case of biosolids (Le 2005; Antille et al. 2008a) or to risk of soil salinisation in the case of compost (Stamatiadis et al. 1999; Kokkora et al. 2008a).

In summary, recycling of compost and biosolids to agricultural soil involves issues associated with varying composition and nutrient imbalances. This makes it difficult to match crop requirements, avoid undesirable effects on the environment and avoid erratic agronomic performance. Although governments and the public generally support the agricultural recycling of compost and biosolids, the farmers face the problem of how to manage these organic materials more efficiently. This chapter addresses these issues by focusing upon some of the aspects that need to be taken into consideration when applying compost and biosolids to agricultural soils in a sustainable manner.

13.2 Using Compost in Agricultural Land

13.2.1 Introduction

Composting is a natural process which involves the aerobic biological decomposition of biodegradable materials under controlled conditions (Misra et al. 2003). The Landfill Directive (CEC 1999) defines biodegradable waste as *any waste that is capable of undergoing anaerobic or aerobic decomposition, such as food and garden waste, and paper and cardboard*. Municipal waste is defined as *waste from households, as well as other waste which, because of its nature or composition is similar to waste from household* (CEC 1999).

Consequently, the term BMSW includes in fact a wide variety of biodegradable materials, including waste as crop residues, household food waste, catering waste or food industry waste (e.g. food processing by-products). The variability in the feedstock material (waste to be composted) results in differences in the composting method that is appropriate for the production of a safe product (compost) for agricultural use. BMSW compost, therefore, is a significantly variable material, in terms of nutrient composition or other physicochemical properties.

13.2.2 Compost Composition

Table 13.1 indicates the variation in nutrient and organic matter status of two types of BMSW compost. The biowaste compost was produced in a centralised plant receiving source-separated BMSW, containing predominantly kitchen, garden,

Table 13.1 Range of dry matter (DM), organic matter (OM) and nutrient status of bio- and vegetable-waste compost

Properties (% DM)	Biowaste compost	Vegetable-waste compost
DM[a]	45.0–74.4	24.7–52.9
OM	42.8–75.8	17.8–48.6
Total N	1.25–2.27	0.64–1.54
C:N	11.7–33.1	9.9–25.4
Soluble mineral N	0.00–0.06	0.00–0.13
Soluble P	<0.01–0.06	<0.01–0.06
Soluble K	0.64–0.96	0.58–1.70

[a]All properties are quoted on dry weight basis, except for DM which is on fresh weight basis

paper and cardboard waste. The vegetable-waste compost was produced from onion waste mixed with straw on farm. The feedstock materials used for the production of these two compost types were suitable for producing high quality compost: source-separated green and kitchen waste (Crowe et al. 2002). The data presented in Table 13.1 were produced from the analysis of five samples per compost type, with a minimum of three replicates per sample, collected in spring of two consecutive years (2005–2006) from two different composting facilities in England.

13.2.2.1 Compost Dry Matter and Organic Matter Content

Compost dry matter (DM) content varied widely between the composts from 25% to 74% fresh weight. This variation was due to the different input materials and composting process used; biowaste compost was produced by in-vessel composting, whereas vegetable-waste compost by static pile composting in uncovered heaps. Differences in DM imply management issues for using compost in agricultural practise. Low DM suggests potential odour problems, and also difficulties in compost application to land due to the bulk of the material, which can result in uneven landspreading. On the other hand, high DM indicates a dusty material which is irritating to work with (MUE 2005). Optimum values of compost DM range from 40% to 60% (Bary et al. 2002). However, compost DM content outside this range is often the case. Studies on various types of BMSW compost reported compost DM in the range of 20–97% (Nevens and Reheul 2003; Tang et al. 2003; Hartl and Erhart 2005).

This variability in compost DM significantly interferes with the agricultural use of compost, as the moisture content of the compost determines the amount of nutrients which will be supplied with a given amount of compost, or in other words, the amount of material necessary for the application of a certain amount of nutrient(s).

Percentage organic matter (OM) also varied widely from 18% to 76% dry weight. The vegetable compost OM was particularly low, which was due to the amount of soil included as a result of the production method used. Other work has demonstrated that compost OM may range from 22% to 77% (Canet et al. 2000; Zmora-Nahum et al. 2007). In general, the OM of all composts is high compared to the soil OM, indicating that composts could be applied to land as soil conditioner to increase soil OM content (He et al. 1992).

13.2.2.2 Compost Nutrient Content (N, P and K)

Compost total nitrogen (N) content ranged between 0.6% and 2.2% dry weight (Table 13.1). Typical values of compost total N vary from 0.8% to 3% (Iglesias-Jimenez and Alvarez 1993; Wolkowski 2003; Zmora-Nahum et al. 2007). Total N content is an important compost property, as the amount of compost land application is often restricted according to the amount of total N applied to the soil.

As mentioned earlier, compost application in the UK is generally limited to a maximum of 250 kg total N ha^{-1} per year. Such practical rates of application aim predominantly not to overload the system with organic N, but also result in controlling the amount of nutrients applied with organic materials. If aiming to apply a certain amount of total N ha^{-1}, the total N content of the fresh compost determines the total amount of fresh material to be added to the soil. For example, using compost with 40% DM and 1.1% total N dry weight (values close to the case of vegetable-waste compost given in Table 13.1), in order to apply 250 kg total N ha^{-1}, a total amount of approximately 57 Mg ha^{-1} of fresh material is necessary. On the other hand, applying 250 kg total N ha^{-1} of a compost with 60% DM and 1.8% total N dry weight (values close to the biowaste compost in Table 13.1), a total amount of 23 Mg ha^{-1} fresh material is necessary, which is less than half the amount needed in the first case. These differences result in different costs of compost use in agriculture, as the higher the amount of compost to be spread, the higher the costs involved per ha (Kokkora 2008). Only a small portion of the compost total N is in mineral forms (Table 13.1). The majority of compost N is bound in organic forms and thus is not directly available to plants. Knowledge of N availability following compost land application is very important for its sustainable use in agriculture. Compost N availability is discussed separately in Sect. 13.2.3.

The concentration of mineral forms of phosphorus (P) was low for both composts, as presented in Table 13.1. Consequently, compost soil application results in the addition of limited amounts of readily available P, thus indicating limited direct replacement potential of fertiliser P. Further contribution in mineral P following compost soil application, however, is expected with the mineralisation of compost OM. A review on the effect of BMSW compost on soil P indicated that it can effectively supply P to the soil, and that increased rates of compost increase plant P uptake (Hargreaves et al. 2008). Soil P availability increases with the addition of compost; however, soil P retention often decreases with increased compost application rates due to increased P mobility and competition for sites on metallic oxides (McGechan and Lewis 2002; Hargreaves et al. 2008). Compost application at high rates to meet N requirements may result in downward movement of P (Bar-Tal et al. 2004). Increased P solubility and downward movement was also observed when biowaste compost was applied at practical rates on coarse sandy soil (Kokkora 2008). Under the same sandy soil conditions, however, vegetable waste compost did not increase P movement downwards through the soil profile. These findings suggest that compost P leaching potential depends on the feedstock material of the compost, the application rates used and also the soil conditions. Phosphorus leaching constitutes a significant environmental concern as it contributes to the eutrophication of freshwater.

Compost soluble potassium (K) content was at higher levels than N and P. In fact, soluble K was at a similar order of magnitude to the total N content of the materials. Hence compost application aiming at 250 kg total N ha^{-1} results in the addition of similar amounts of K, suggesting the compost potential for replacing significant amounts of mineral fertiliser K. In general K availability in compost is similar to mineral K fertilisers (Hargreaves et al. 2008). Research has shown that

Table 13.2 Heavy metal content of bio- and vegetable-waste compost

Determinant ($mg\ kg^{-1}$ DM)	Biowaste compost	Vegetable-waste compost	BSI PAS 100: 2005
Cd	0.4–0.9	0.9–1.1	1.5
Cr	25–32	57–63	100
Cu	56–79	29–41	200
Pb	50–82	11–22	200
Hg	<0.1–0.1	<0.1–0.1	1
Ni	15–23	16–17	50
Zn	176–177	56–74	400

BMSW compost application increases the levels of extractable K on silt loam and silty clay loam soils (Parkinson et al. 1999; Hartl et al. 2003). Kokkora et al. (2008b) demonstrated that the extractable K content of sandy loam soil increased linearly with BMSW compost application, in response to the increase in the amount of compost soluble K applied. These findings indicate that compost application at practical rates potentially reduces or replaces the need for mineral fertiliser K.

13.2.2.3 Heavy Metals

The heavy metal content of the bio- and vegetable-waste composts is presented in Table 13.2, along with the maximum allowable concentrations of heavy metals according to BSI PAS 100 (BSI 2005). The data shown were produced from the analysis of two samples per compost, using three replicates per sample. Both composts analysed in both years were well within the BSI PAS 100 limits for heavy metal contaminants. Hence, both composts can be applied to agricultural soil without posing an environmental risk from soil contamination with heavy metals. In general, compost produced from source-separated BMSW is likely to be low in heavy metals and within the existing limits for agricultural use. On the contrary, compost produced from non-separated MSW or non-source-separated BMSW is often problematic regarding its heavy metal content (Gomez 1998; Brinton 2000).

13.2.3 Nitrogen Release

Knowledge of the short- and long-term availability of N following the application of compost is essential in order to meet crop requirements, whilst ensuring environmental protection from increased nitrate leaching. In general, N availability following compost application is low, because the majority of the total compost N is bound to the organic N pool (Amlinger et al. 2003; Hartl and Erhart 2005). The mineral N content of compost is low (as shown in Table 13.1) because N is partly lost during composting due to volatilisation (Zwart 2003). The organic portion of compost total N is not readily available to plants; it can be mineralised, and then potentially taken up by the plants, immobilised, denitrified, volatilised, fixed within the clay minerals and/or leached. The main losses of N from the soil are due to crop removal and leaching (Tisdale et al. 1999).

13.2.3.1 Compost N Mineralisation

The mineralisation of compost N involves the conversion of organic forms of N to ammonia (NH_3) or ammonium (NH_4^+) and nitrate (NO_3^-). Nitrogen mineralisation always occurs simultaneously with N immobilisation in the soil, which operates in the reverse direction, with the soil microbial biomass assimilating inorganic N forms. The relative magnitudes of N mineralisation and immobilisation determine whether the overall effect is net mineralisation or net immobilisation (Cabrera et al. 2005).

Whether N is mineralised or immobilised depends primarily on the C:N ratio of the material undergoing decomposition by the soil microorganisms (Benbi and Richter 2003). The contribution of compost organic N towards the N mineralised following the application of different types of BMSW compost, with varying C:N ratios, to a wide variety of soils, including poor acidic, clay, sandy and sandy loam soils, ranged from negative (immobilisation) to lower than 12% (Beloso et al. 1993; Gagnon and Simard 1999; Mamo et al. 1999; Kokkora 2008). Figure 13.1 presents the nitrogen mineralisation rate (NMR) as a percentage of the total and organic compost N applied, as reported from lysimeter, pot and incubation trials involving BMSW composts application to light soils (adapted from Kokkora 2008).

BMSW compost with C:N ratio less than 15 is likely to result in net mineralisation of organic N, rather than immobilisation (Kokkora 2008), which is also the case for other types of organic materials (Iglesias-Jimenez and Alvarez 1993; Chadwick et al. 2000; Gutser et al. 2005). Compost with C:N ratio higher than 15 is

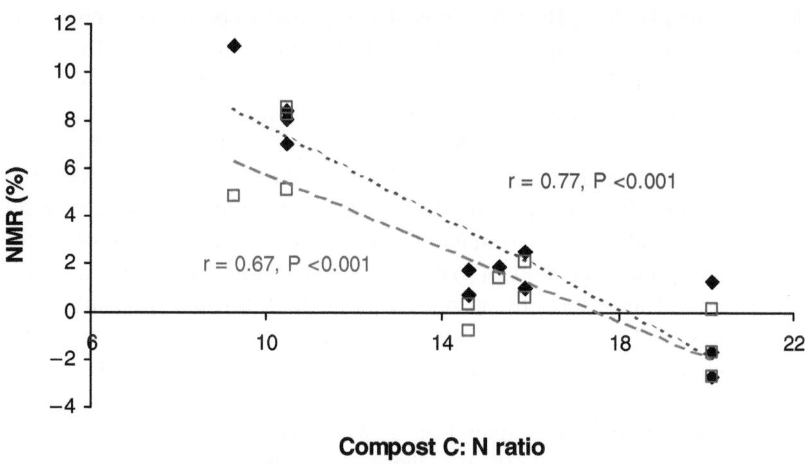

Fig. 13.1 Compost N mineralisation in relation to compost C:N ratio (*filled diamond*: as a percentage of the total N applied, and *open square*: as a percentage of the organic N applied)

likely to result in limited N mineralisation or immobilisation of organic N, depending on compost C quality and soil and climatic conditions, the increase in compost C: N ratio favouring N immobilisation. Studies on N mineralisation from various organic materials have shown that the decomposability of the carbon sources of the materials also influences N mineralisation: the more recalcitrant the carbon source to decomposition, the slower the N mineralisation (Rogers et al. 2001). Furthermore, studies monitoring the N mineralisation with time provided evidence of a multiple-phase mineralisation following compost application to the soil, which was due to the different decomposability of the carbon sources (He et al. 2000; Rogers et al. 2001; Kokkora 2008), thus suggesting a non-steady N release, even if a compost is expected to result in net N mineralisation (C:N <15). This finding is important because it implies a non-continuous or increasing N availability in the soil solution which would potentially increase the N leaching losses, but a slow release of N with time which can increase the N use efficiency by the crop.

In comparison to sewage sludge, biosolids, manures, or other non-composted organic amendments, compost application results in lower N mineralisation rates (Petersen 2003; Gutser et al. 2005). Han et al. (2004) demonstrated that a combined application of compost with chemical fertiliser could improve the compost use efficiency by increasing mineralisation of compost N, especially in soils with low mineral N content.

13.2.3.2 Nitrogen Uptake

Nitrogen availability is the main factor determining crop production. In different studies, crop N recovery ranged between 0% and 15% of the total BMSW compost N applied (Mamo et al. 1999; Amlinger et al. 2003; Nevens and Reheul 2003; Wolkowski 2003; Hartl and Erhart 2005; Kokkora 2008).

Nitrogen uptake from BMSW compost is reported to be largely dependent on compost C:N ratio and application rate used. Increased compost C:N resulted in decreased N uptake due to decreased N mineralisation (Iglesias-Jimenez and Alvarez 1993; Eriksen et al. 1999; Kokkora 2008). According to Kokkora (2008) the application of compost with C:N of 20 to sandy soil at rates of 400 kg total N ha^{-1} resulted in lower N uptake during the early stages of forage maize growth, than the non-amended soil, because of early season immobilisation of compost and soil N. Eriksen et al. (1999) also reported an early season immobilisation of soil N following the application of about 310 kg total N ha^{-1} of compost with a C:N ratio of 40. Rogers et al. (2001) found decreased NMR following the application of 224 kg total N ha^{-1} of food processing waste with a C:N of 31 in comparison to the NMR at half the rate. They suggested that the higher application rate may have overloaded the soil ecosystem, exceeding the maximum rate at which the soil microbial community could mineralise N, whilst possibly other factors, such as excess salts, were negatively affecting the soil microbes.

Crop N uptake from compost-amended soil is dependent on crop N requirements, soil available N and N mineralised from the compost. In most cases, due to the low N mineralisation potential of the compost, compost application at practical rates is not adequate to reach the crop maximum N uptake (Iglesias-Jimenez & Alvarez 1993; Eriksen et al. 1999; Petersen 2003; Kokkora 2008). Sullivan et al. (2002) showed that BMSW compost application resulted in reduction in N fertiliser requirement during midseason tall fescue growth at levels equivalent to 3.6–5.2% of compost N in the second year, and 2.0–3.5% of compost N in the third year following application. Hence, supplementary application of mineral fertiliser N might be necessary. On the other hand, there have been cases where due to the high N mineralisation potential of the soil, crop N uptake was slightly or not affected by compost application (Hartl et al. 2003; Kokkora 2008). These findings underline the importance of a careful management of compost use in agriculture in order to result in optimum crop production. Compost management, however, needs to ensure minimum environmental risks such as nitrate leaching.

13.2.3.3 Nitrogen Leaching from Compost-Amended Soil

Leaching is the process of N losses by the downward movement of water through the soil profile (Benbi and Richter 2003). Nitrate leaching is generally a major N loss mechanism from agricultural soils in humid climates and under irrigated cropping systems (Tisdale et al. 1999). Research has shown that BMSW compost application at practical rates does not substantially increase nitrate leaching. Hartl and Erhart (2005) applied various rates of compost (C:N ranged between 9 and 49) over a 10-year field experiment on silty clay loam soil. They showed that on average of the 10 years the soil NO_3–N content (0–0.9 m depths) increased from 6 to 14 kg ha^{-1}, corresponding to the application of about 81–205 kg total N ha^{-1} per year. Similar results were reported by Nevens and Reheul (2003) for a 4-year compost application (C:N of 10.2 at about 335 kg total N ha^{-1} per year). Kokkora (2008) in a 1-year lysimeter trial reported that the maximum increase in the amount of mineral N leached from compost-amended sandy soil was 3.8 kg N ha^{-1} (C:N ranged from 11 to 20, and application rates from 100 to 600 kg total N ha^{-1}).

Work involving the application of high rates of compost N, however, has shown higher potential for mineral N leaching following compost application. Eriksen et al. (1999), demonstrated that application of compost with C:N of 40 at about 926 kg total N ha^{-1} resulted in significantly higher residual soil NO_3–N compared to the control sandy soil the second year after application. Mamo et al. (1999) showed that the application of compost, with C:N ratio ranging from 15 to 27, to loamy sand soil at high rates (about 900 kg total N ha^{-1} annually) increased the NO_3–N leaching 1.4–2.6-fold compared with the non-amended soil over a 3-year period. In general, compost application at practical rates was shown not to increase substantially the NO_3–N leaching losses on light textured soils, even under irrigation and high drainage rate conditions. This finding is generally related to the slow compost N mineralisation, which leads to more efficient uptake by the growing crop.

In addition, the majority of existing research has demonstrated an increase in the soil organic N content following compost application. Soil storage of organic N in the compost-amended soil may explain the limited NO_3–N leaching despite the low crop N recovery. Hartl and Erhart (2005) observed that the increase in soil organic N following compost application was complemented with an increase in organic C within the top 0.3 m soil depths. They concluded that this increase suggests that organic N is tied up in the SOM, hence supporting the low NO_3–N leaching following compost application.

13.3 Recycling of Sewage Sludge (Biosolids)

13.3.1 Introduction

Sewage-sludge is defined in Article 2(a) of the EU Sewage Sludge Directive 86/278/ECC (EC 1986) as: *The residual sludge from sewage plants treating domestic or urban wastewaters and from other sewage plants treating wastewaters of a composition similar to domestic and urban wastewaters.*

Treated sludges are usually referred to as biosolids (MAFF 2000). These materials have been treated to achieve certain standards that make them suitable for their use on agricultural land. These, along with some guidelines for their application, are specified in "The Safe Sludge Matrix" (Chambers 2001). Under these specifications, only enhanced treated sludge (i.e. $\geq$99.9999% pathogens destruction) can be applied to all crops and grassland.

13.3.2 Nutrient Content of Sewage Sludge (Biosolids)

The fertiliser value of sewage sludge (biosolids) is directly related to the nutrient content of the material, particularly nitrogen and phosphorus (CIWEM 1995). The content of nitrogen and phosphorus is largely dependent on the treatment process and the type of sludge (liquid, cake or dried). Their concentration is usually sufficient to cover the requirements of agricultural crops (Gurjar 2001) and to maintain soil fertility (MAFF 2000). The content of potassium is often very low (Gurjar 2001) and is not enough to meet crop requirements when the products are applied at rates in line with standard farm practice. Table 13.3 shows the nutrient composition of thermally dried biosolids ($\geq$90% DM) in a number of samples taken from a wastewater treatment works in the UK.

13.3.2.1 Nitrogen

There are a number of factors controlling the availability of nitrogen (N) in biosolids. Because the conversion of organic-N into ammonium-N is a slow process (Black 1968), sludges are often regarded as slow-release fertilisers (Le 2005). Prior

Table 13.3 Chemical composition of thermally dried biosolids (≥90% DM) for a wastewater treatment work in the UK

Determination	Value	Units
Total carbon (C)	31.10	% [w/w]
Total nitrogen (N)	3.89	% [w/w]
Soluble nitrogen (N)	Trace	–
C:N ratio	8.00	–
Total phosphorus (P_2O_5)	5.73	% [w/w]
Water-soluble phosphorus (P_2O_5)	0.19	% [w/w]
N:P ratio	0.68	–
Water-soluble potassium (K_2O)	0.14	% [w/w]
Total sulphur (SO_3)	2.95	% [w/w]
Total calcium (Ca)	2.97	% [w/w]
Total magnesium (Mg)	0.25	% [w/w]

Note: Some deviation (±30%) of these data may be expected for individual samples

to treatment, the majority of the nitrogen is organic and mostly insoluble. During the anaerobic digestion, up to 50% of the organic-N is transformed to ammoniacal-N which is soluble (CIWEM 1995). Hence, the digestion process changes the quality of the sludge treated, affecting its fertiliser value (Demuynck et al. 1984). The anaerobic digestion contributes to reducing the C:N ratio of the sludge, since the process consumes carbon. Consequently, the mineralisation rate can increase following application to land. However, if after digestion the sludge undergoes dewatering, the soluble-N is carried in the water leaving behind mainly organic-N fractions (CIWEM 1995) which needs to be mineralised to become available to crops. Thus, the N:P ratio in the biosolid is affected by digestion followed by dewatering, since the total N content is significantly reduced. This is consistent with the analysis presented in Table 13.3 where majority of the N in the biosolid is organic-N. In this particular case, the resultant N:P ratio is as low as 0.68, thus augmenting the nutrient imbalances in the biosolid. The C:N ratio shown in Table 13.3 is relatively low (<20). Therefore, it could be predicted that mineralisation of organic-N fractions in the biosolid would occur rapidly following application to land. However, mineralisation of nitrogen is also controlled by the degree of stabilisation of the organic matter (Hall 1984), soil temperature and soil moisture content (Sullivan et al. 2007). Accurate estimation of nitrogen availability from biosolids is crucial to recommending application rates that better satisfy crop requirements and reduce N leaching (Cogger et al. 2004). USEPA (1983) reported for biosolids undergoing anaerobic digestion that up to 20% of the organic-N mineralises in the first year of application. In the second year this figure decreases to 10%. Le and Gedara (2007) reported that residual organic-N mineralises at a rate of 7% or lower in subsequent years. MAFF (2000) indicates that nitrogen availability in the first year of application for digested cake (≈25% DM) and thermally dried biosolids (≈90% DM) applied in spring in the UK is approximately 15%. Further work conducted by Cogger et al. (2004) with thermally dried biosolids indicated availabilities between 28% and 44% in the first year and between 5% and 8% in the second year. These values suggest that there is a residual effect of biosolids N over a long period of time and that mineralisation continues at low rate in subsequent crops. Organic-N in the biosolid that is not initially mineralised must be accounted for in the following biosolids applications.

Therefore, when biosolids are regularly applied to the same field, the application rate required to meet the demand of the crops may be reduced (Sullivan et al. 2007). Gibbs et al. (2007) studied the effects of the application of thermally dried biosolids on N leaching using winter wheat (*Triticum aestivum L*). According to their findings, N leaching for a medium textured soil which received 250 kg [N] ha^{-1} was increased by approximately 4% of the total N applied compared with untreated soil. However, N leaching in both treated and untreated soils was statistically similar. The same work reported N utilisation efficiencies grain in the range 10–18% of total N compared with mineral fertiliser. Corrêa (2004) found that the efficiencies of biosolids in supplying N to rye-grass were similar to that of mineral fertilisers in a spodsol soil, whereas for an oxisol soil they were between 65% and 67%. These results indicate that the capacity of biosolids to supply N depends upon the crop, the biosolids' characteristics, and the soil type.

13.3.2.2 Phosphorus

The majority of the phosphorus in biosolids is inorganic and it is present in the form of iron, aluminium and calcium phosphates (Wise 1999). A small amount of organic-P (usually less than 6% of the total-P) in the form of organic phosphates may also be found (EC 2003). The formation of inorganic compounds takes place in the tertiary treatment during the chemical removal of P from the wastewater effluent by means of precipitation. Withers and Flynn (2006) reported that enhanced biological phosphate removal can also result in rich-P biosolids. Research has shown that the largest phosphorus content in sludges can occur when enhanced removal by precipitation with iron, aluminium and calcium is performed (Wise 1999). This can result in sludges having up to 11% P as P_2O_5 (Le and Gedara 2007). Removal by precipitation also affects the bioavailability of P since insoluble inorganic-P compounds are not readily available for plant uptake (de Haan 1980; Lu and O'Connor 2001). This information appears to be consistent with that shown in Table 13.3 where less than 3.5% of the total P is water-soluble-P. AWA (1979) reported that P availability in the first year for biosolids containing iron or aluminium salts is significantly lower than 50%. Maguire et al. (2001) found that the amount of extractable P (Olsen's-P) in soil was higher when the biosolids applied were produced without iron and aluminium salts or lime.

Anaerobic digestion has been reported to increase the quantities of available phosphorus. The digestion process breaks down organic compounds in the sludge thereby releasing phosphate ions (Wise 1999). A major concern arising from the recycling of biosolids to agricultural land is the inherent risk of building up soil P levels. In England and Wales, Skinner et al. (1992) showed that an important proportion of arable soils (22%) had P-indexes in excess of 3 ($\geq$46 mg L^{-1} [Olsen's-P]); whereas for grassland up to 30% of the soils had P-indexes

3 (26–46 mg L^{-1} [Olsen's-P]) or lower. The implication of these findings is that the fertility status of the soils with particular regards to P can potentially reduce the land-bank for recycling. It is well known that any surplus of P in the soil provides the opportunity for P losses. Hence, the application of biosolids containing high levels of P can have environmental implications when biosolids are applied to soils with high P-indexes. On the other hand, rich-P biosolids can be a useful, and certainly inexpensive, source of P in cases where continuous cultivation has run down soil-P below the level needed for optimum crop yield. Johnston and Syers (2006) suggested a critical level of plant available soil-P which is unique to the soil type and the cropping system. This level should always be maintained to maximise the efficiency at which soil-P is taken up by crops and ensure soil-P is not unnecessarily built up (Johnston and Syers 2006).

Lu and O'Connor (2001) found that iron- and aluminium-rich biosolids are expected to act as slow release P-sources, thereby being less prone to leaching than soluble P sources. Withers and Flynn (2006) highlighted that P in biosolids, despite having low water solubility, can raise soil-P tests (e.g. Olsen's-P) to a greater extent than might be predicted from their solubility in water. Due to the low N:P ratio of biosolids, application at rates based only on crop requirements for P would result in N being largely under-dosed. On the other hand, application rates based on N can increase the soil-P index in the long term when inorganic compounds become progressively available. Maguire et al. (2001) argued that this effect may be mitigated by the simultaneous increase of iron and aluminium levels in the soil. A number of studies (Holford et al. 1997; Hooda et al. 2000; Withers and Flynn 2006) led to the conclusion that once the percentage P saturation capacity in the soil is in excess of 25%, P diffuses more easily into the soil solution, increasing the risk of losses. Knowledge of P content in the biosolids and P status in the soil could prevent undesirable effects on the environment, e.g. eutrophication of water courses.

In cereal crops and oilseed rape (*Brassica napus L*), application of P can be significantly reduced or even avoided when soil P-index is above 3 ($\geq$46 mg L^{-1} [Olsen's-P]) (MAFF 2000). In most cases, recovery of sludge-P in the crop is lower than that of mineral fertilisers although for both materials the recovery of P is usually low, i.e. <10% (Withers and Flynn 2006). Previous work by Corrêa (2004) with rye-grass indicated values of P recovery from biosolids in the range 1.2% to 7%. MAFF (2000) reported that up to 50% of the total phosphate content of digested cake and thermally dried biosolids is available to the next crop grown, with the remaining 50% being available in subsequent years.

13.3.2.3 Potassium

The content of potassium (K) in sewage sludge (biosolids) is very small (Table 13.3). It is essentially all water-soluble in the form of chloride, sulphate and carbonate salts. The low K content generally encountered in biosolids significantly increases the nutrient imbalances in terms of N:P:K ratios. The standard farm practice is to apply K fertilisers independently from biosolids if necessary.

Table 13.4 Proposed EU limits and heavy metals composition of thermally dried biosolids (≥90% DM) for a wastewater treatment work in the UK

Heavy metal	Value	EU Limit	Unit
Total cadmium (Cd)	1.2	10	mg kg^{-1}
Total zinc (Zn)	493	2500	mg kg^{-1}
Total copper (Cu)	329	1000	mg kg^{-1}
Total chromium (Cr)	70.9	1000	mg kg^{-1}
Total mercury (Hg)	1.1	10	mg kg^{-1}
Total nickel (Ni)	34.4	300	mg kg^{-1}
Total lead (Pb)	163.5	750	mg kg^{-1}
Total molybdenum (Mo)	9.1	–	mg kg^{-1}
Total selenium (Se)	2.9	–	mg kg^{-1}

Note: Some deviation (±20%) of these data may be expected for individual samples

13.3.2.4 Heavy Metals

Heavy metals in sewage sludge (biosolids) originate from a number of different sources such as industrial and domestic household waste, eroding pipes, and runoff from roads. In recent years the content of heavy metals in sewage sludge has decreased significantly as a result of pre-treatment of industrial waste (Le 2005). Additionally, the use of enhanced sludge treatment methods, such as acid and alkaline thermal hydrolysis, can reduce the heavy metal content in sludge cakes (Dewil et al. 2006). Some heavy metals applied with biosolids provide a source of micronutrients, especially in deficient soils. However, their concentration in the soil must not exceed specified limits set out in The Sludge (Use in Agriculture) Regulations, to ensure that health risks to stock and humans are minimised. As heavy metals are relatively immobile, they tend to accumulate in the top soil. Soil pH is the main factor affecting the availability of heavy metals to plants. This decreases markedly when soil pH increases above neutrality, except for selenium and molybdenum (Darwich 1998). Although most sludges (biosolids) are considered to be fit for land application, recycling cannot be practised when soil pH is 5 or lower. Table 13.4 gives an indication of some heavy metals present in biosolids together with the EU limits as proposed in the 3rd Draft EU Sewage Sludge Directive (EC 2000).

The suitability of this particular biosolid for spreading on agricultural land will also depend on specific soil parameters, e.g. pH and existing heavy metal levels in the soil.

13.3.3 Nutrient-Enriched Biosolids

13.3.3.1 Introduction

In the previous sections, some of the factors affecting the agricultural recycling of sewage sludge (biosolids) were examined. There is a need to increase current levels of biosolids use by farmers to deal with increasing production. However, the fertiliser value of biosolids, as defined by their nutrient composition, nutrient concentration and release characteristics, remains a major concern. The following section discusses how the fertiliser value of biosolids can be enhanced by improving their nutrient composition.

13.3.3.2 A Sustainable Solution for the Recycling of Biosolids to Land

Mineral fertilisers such as urea (46% N) and muriate of potash (60% K_2O) can be used to provide more suitable N:P:K ratios and improve biosolids' formulation. The addition of mineral fertilisers to produce nutrient-enriched biosolids, namely organomineral fertilisers (OMFs), addresses to certain extent the problems associated with the fertiliser value of biosolids. Antille et al. (2008b) reported that the composition of biosolids having 3:4(+):0.15 ($N:P_2O_5:K_2O$) was modified to obtain more convenient ratios by adding urea and muriate of potash. The resultant compositions of the OMFs were 15:4:4 and 10:4:4. Following application to land, the mineral fraction of the OMF provides plant nutrients which are readily available for plant uptake, thus supplying the demand for the main growth period of the crop. The organic fraction (biosolid) provides a slower release source of nutrients which become available throughout the growing season. This component can help to reduce the need for fertiliser applications later in the season, so that further nitrogen dressings can be performed by adjusting the demand of the crop with the organic OMF-N being mineralised. The agronomic characteristics of a number of OMF products is being investigated (Antille et al. 2008a, b).

The development of OMFs introduces a sustainable approach towards recycling of biosolids to agricultural land (Gedara et al. 2008).

It provides farmers with more reliable products in terms of product formulation, physical characteristics, and nutrient supply compared with biosolids. The use of OMFs may result in increased confidence in the use of organic materials by farmers and may also contribute to reducing reliance on mineral fertilisers which are constantly increasing in price.

13.4 Conclusions

The recycling of compost and biosolids to agricultural land appears to be a sustainable option. Increased rates of recycling may reduce reliance on mineral fertilisers, which are increasing in price. However, the recycling of organic materials to agricultural land requires a good understanding of the interactions between materials' characteristics and the soil–crop system, and the effects on the environment.

The fertiliser value of compost and biosolids is determined by their nutrient content and availability, especially with regards to nitrogen, phosphorus and potassium. Compost and biosolids are regarded as slow-release fertilisers. Nitrogen availability following application of biosolids is expected to be between 15% and 20% (1st year), about 10% (2nd year), and 7% or less thereafter. For composts, nitrogen availability is also low. Compost application results in nitrogen mineralisation varying from negative (immobilisation) to 12%, depending largely on the compost's C:N ratio.

The recovery of nitrogen from biosolids is expected to be approximately 15% for cereals and between 6% and 75% for grassland. Nitrogen recovery from compost ranged between 0% and 15% in a wide variety of soils and crops (cereals and grass). Generally, application of composts and biosolids at sensible rates, i.e. not exceeding crop requirement, does not substantially increase nitrogen leaching. This is due to the slow release of nitrogen which can be progressively taken up by the growing crop. However, nitrogen leaching does take place when there is a mismatch between crop uptake and availability of nitrogen. Continuous site application of P-rich biosolids can raise soil extractable-P in the long term. Up to 50% of the total biosolids-P applied can become available within the year of application. P availability in compost is lower than in biosolids. Soil P availability increases with compost application; however, fertiliser P addition may be necessary in low P-index soils. Application of compost increases the level of soil extractable-K. Compost application at practical rates can reduce substantially the need for potassium fertilisers.

However, in biosolids, the content of potassium is negligible and application of potash may be needed in low K-index soils.

Heavy metals in composts produced from source-separated BMSW are low and well within the maximum allowable limits; hence their use in agriculture is safe from a heavy metal contamination perspective. A pH value of 5 is the cut-off level for any application of biosolids to land due to the risk associated with heavy metal availability.

The development of nutrient-enriched biosolids (organomineral fertilisers) appears to be a sustainable alternative for the recycling of biosolids to land. It enhances the nutrient value of biosolids by correcting imbalances in biosolids' nutrient composition and their nitrogen release characteristics. Development of potential techniques for compost nitrogen enrichment is recommended for future research.

Acknowledgements The authors are grateful to United Utilities plc, EPSRC (Engineering and Physical Sciences Research Council, UK), Korgialenion Athlon Foundation (Greece), and staff at the School of Applied Sciences at Cranfield University (UK).

References

Amlinger F, Götz B, Dreher P, Geszti J, Weissteiner C (2003) Nitrogen in biowaste and yard waste compost: dynamics of mobilisation and availability - a review. Eur J Soil Biol 39:107–116

Antille DL, Sakrabani R, Godwin RJ (2008a) Organomineral fertilisers: nitrogen dynamics and evaluation of agronomic characteristics. Proceedings of ASABE (American Society of

Agricultural and Biological Engineers) Annual International Meeting, 29 June–2 July 2008, Providence, US, ASABE (Paper No: 084000)

Antille DL, Sakrabani R, Godwin RJ (2008b) Methodological approach to identifying the properties of a novel organomineral fertiliser: crop responses to organomineral fertilisers in the UK and measurement of nitrogen availability in controlled laboratory conditions – Part I: agronomic aspects. Cranfield Multi-Strand Conference: Creating wealth through research and innovation (CMC 2008) (Paper no. 124). Cranfield University, MK43 0AL, UK, 6–7 May 2008

AWA (Anglian Water Authority) (1979) Manual of good practice for utilisation of sewage sludge as a fertiliser. AWARD Service to Agriculture, UK

Bar-Tal A, Yermiyahu U, Beraud J, Keinan M, Rodenberg R, Zohar D, Rosen V, Fine P (2004) Nitrogen, phosphorus, and potassium uptake by wheat and their distribution following, successive, annual compost applications. J Environ Qual 33(5):1855–1865

Bary A, Cogger C, Sullivan D (2002) What does compost analysis tell you about your compost? Available via: http://www.soils1.org. Cited 2005

Beloso MC, Villar MC, Cabaneiro A, Carballas M, González-Prieto SJ, Carballas T (1993) Carbon and nitrogen mineralization in an acid soil fertilized with composted urban refuses. Bioresource Technol 45:123–129

Benbi DK, Richter J (2003) Nitrogen dynamics. In: Benbi DK, Nieder R (eds) Handbook of processes and modelling in the soil-plant system. Haworth, New York, US, pp 409–482

Black CA (1968) Soil – plant relationships, 2nd edn. Wiley, New York

Boulos S, Pocock R, Gilbert EJ (2006) The state of composting and biological waste treatment in the UK 2004/05. The Composting Association, Wellingborough

Bowden W, Hann MJ (1997) The availability of nitrogen following digested sludge incorporation in arable land. Nutr Cycl Agroecosys: 1–5. Kluwer Academic, The Netherlands. In: Bowden W (1996) The availability of nitrogen following digested sludge incorporation in arable land. PhD thesis, Cranfield University at Silsoe, UK

Brinton WF (2000) Compost quality standards and guidelines. An international view. Final report to New York State Association of Recyclers. Woods End Research Laboratory. http://www.woodsend.org/pdf-files/nysar-ne.pdf. Cited 2004

BSI (British Standards Institution) (2005) Publicly available specification for composted materials PAS 100: 2005. BSI, London, UK

Cabrera ML, Kissel DE, Vigil MF (2005) Nitrogen mineralization from organic residues: research opportunities. J Environ Qual 34:75–79

Canet R, Pomares F, Albiach R, Tarazona F, Ibañez MA, Ingelmo F (2000) Analyzing chemical properties of MSW composts. BioCycle 41:72–76

CEC (Commission of European Communities) (1986) Directive on the protection of the environment, and in particular of the soil, when sewage sludge is used in agriculture (86/278/EEC). Official Journal L187/6, 4 July 1986

CEC (Commission of European Communities) (1999) Council Directive 1999/31/EC of 26 April 1999 on the landfill of waste. Official Journal of the European Communities L182:1–19

Chadwick DR, John F, Pain BF, Chambers BJ, Williams J (2000) Plant uptake of nitrogen from the organic fraction of animal manures: a laboratory experiment. J Agric Sci 134:159–168

Chambers B (2001) The safe sludge matrix: guidelines for the application of sewage sludge to agricultural land, 3rd edn. British Retail Consortium, Water UK, ADAS-Gleadthorpe Research Centre http://www.assuredproduce.co.uk/_code/common/item.asp?id=4033093. Cited 12 September 2008

CIWEM (Chartered Institution of Water and Environmental Management) (1995) Sewage sludge: utilization and disposal. Handbooks of UK Water Practice. CIWEM, London, UK

Cogger CG, Bary AI, Sullivan DM, Myhre EA (2004) Biosolids processing effects on first and second year available nitrogen. Soil Sci Soc Am J 68:162–167

Corrêa RS (2004) Efficiency of five biosolids to supply nitrogen and phosphorous to ryegrass. Pesq Agrope Bras 39(11):1133–1139

Crowe M, Nolan K, Collins C, Carty G, Donlon B, Kristoffersen M, Brogger M, Carlsbaek M, Hummelshoj RM, Thomsen CL (2002) Part 3: Technology and market issues. European Environment Agency (EEA), Copenhagen, DK
Darwich NA (1998) Manual de fertilidad de suelos, 1ra Edición. Instituto Nacional de Tecnología Agropecuaria (INTA) EEA Balcarce – FCA Mar del Plata, Argentina
de Haan S (1980) Sewage sludge as a phosphate fertiliser. In: Hucker TWG, Catroux G (eds) Phosphorus in sewage sludge and animal waste slurries. Proceedings of the EEC Seminar organised jointly by the CEC (Commission of the European Communities) and the Institute for Soil Fertility (Haren, The Netherlands), 12 – 13 June 1980, D Reidel, Groningen, pp 149
DEFRA (Department for Environment Food and Rural Affairs) (2007) e-Digest statistics about waste and recycling: sewage sludge. http://www.defra.gov.uk/environment/statistics/waste/wrsewage.htm. Cited 10 August 2007
Demuynck M, Nyns EJ, Naveau H (1984) Use of digested effluents in agriculture. In: Williams JH, Guidi G, L'hermite P (Eds) Long term effects of sewage sludge and farm slurries applications. Proceedings of a Round-Table Seminar organised by the Commission of European Communities, Directorate General Science, Research and Development, Environment Research Programme, Pisa (Italy), 25–27 September 1984. Elsevier Applied Sciences, London, UK, pp 2
Dewil R, Baeyens J, Neyens E (2006) Reducing the heavy metal content of sewage sludge by advanced sludge treatment methods. Environ Eng Sci 23(6):994–999
EA (Environment Agency) (2004) The state of soils in England and Wales. Report of May 2004, Environment Agency, Bristol, UK
EC (European Commission) (1986) Council Directive EU Sewage Sludge Directive 86/278/EEC of 12 June 1986 on the protection of the environment and the soil when sewage sludge is used in agriculture (OJ L 181, 4.7.1986, 6 p. Directive as last amended by the 1994 Act of Accession)
EC (European Commission) (2000) Working document on sludge. 3rd Draft [ENV.E3/LM] http://ec.europa.eu/environment/waste/sludge/pdf/sludge_en.pdf. Cited 18 September 2008
EC (European Commission) (2003) Draft discussion document for the ad-hoc meeting on bio-wastes and sludges. Directorate General Environment. Directorate A – Communications, governance, production, consumption, and waste. ENV.A2 – Production, consumption, and waste. DG ENV. A.2/LM, Brussels, 15–16 January 2004
EC (European Commission) (2007) Biodegradable waste. http://ec.europa.eu/environment/waste/compost/index.htm, Cited 2007
Eriksen GN, Coale FJ, Bollero GA (1999) Soil nitrogen dynamics and maize production in municipal solid waste amended soil. Agron J 91:1009–1016
Gagnon B, Simard RR (1999) Nitrogen and phosphorus release from on-farm and industrial composts. Can J Soil Sci 79:481–489
Gedara S, Le MS, Antille DL, Tyrrel SF, Sakrabani R, Godwin RJ (2008) Organomineral fertilisers for sustainable sludge recycling. Proceedings of 13th European Biosolids and Organic Resources Conference, Workshop, and Exhibition organised by Aqua-Enviro Technology Transfer, 10 – 12 November 2008, Manchester, UK. http://www.european-biosolids.com/proceedings.php. Cited 20 September 2008
Gibbs P, Chambers B, Benett G, Smith R, Boyle H (2007) Nitrogen release properties of thermally dried and plate-pressed vacuum dried biosolids. In: Lowe P, Horan NJ (2007) (Eds) Proceedings of the 12th European Biosolids and Organic Resources Conference, Workshop, and Exhibition organised by Aqua-Enviro Technology Transfer, 12–14 November 2007, Manchester, UK. http://www.european-biosolids.com/proceedings.php. Cited 20 September 2008
Gomez A (1998) The evaluation of compost quality. Trends Anal Chem 17(5):310–314
Gurjar BR (2001) Sludge treatment and disposal. Balkema, The Netherlands
Gutser R, Ebertseder T, Weber A, Schraml M, Schmidhalter U (2005) Short term and residual availability of nitrogen after long term application of organic fertilisers on arable land. J Plant Nutr Soil Sci 168:439–446

Hall JE (1984) The cumulative and residual effects of sewage sludge nitrogen on crop growth. In: Williams JH, Guidi G, L'hermite P (eds) Long term effects of sewage sludge and farm slurries applications. Proceedings of a Round-Table Seminar organised by the Commission of European Communities, Directorate General Science, Research and Development, Environment Research Programme, Pisa (Italy), 25 – 27 September 1984. Elsevier Applied Sciences Publishers, London, UK, pp 73

Han KH, Choi WJ, Han GH, Yun SI, Yoo SH, Ro HM (2004) Urea-nitrogen transformation and compost nitrogen mineralization in three different soils as affected by the interaction between both nitrogen inputs. Biol Fertil Soils 39:193–199

Hargreaves JC, Adl MS, Warman PR (2008) A review of the use of composted municipal solid waste in agriculture. Agric Ecosyst Environ 123:1–14

Hartl W, Erhart E (2005) Crop nitrogen recovery and soil nitrogen dynamics in a 10-year field experiment with biowaste compost. J Plant Nutr Soil Sci 168:781–788

Hartl W, Putz B, Erhart E (2003) Influence of rates and timing of biowaste compost application on rye yield and soil nitrate levels. Eur J Soil Biol 39:129–139

He X-T, Traina SJ, Logan TJ (1992) Chemical properties of municipal solid waste composts. J Environ Qual 21:318–329

He ZL, Alva AK, Yan P, Li YC, Calvert DVSPJ, Banks DJ (2000) Nitrogen mineralization and transformation from composts and biosolids during field incubation in a sandy soil. Soil Sci 165:161–169

Hogg D, Barth J, Favoino E, Centemero M, Caimi V, Amlinger F, Devliegher W, Brinton W, Antler S (2002) Comparison of compost standards within the EU, North America and Australia. Waste & Resources Action Programme (WRAP), Oxon, UK

Holford ICR, Hird C, Lawrie R (1997) Effects of animal effluents on the phosphorus sorption characteristics of soils. Aust J Soil Res 35:365–373

Hooda PS, Rendell AR, Edwards AC, Withers PJA, Aitken MN, Truesdale VW (2000) Relating soil phosphorus indices to potential phosphorus release to water. J Environ Qual 29:1166–1171

Iglesias-Jimenez E, Alvarez CE (1993) Apparent availability of nitrogen in composted municipal refuse. Biol Fertil Soils 16:313–318

Johnston AE, Syers JK (2006) Changes in understanding the behaviour of soil and fertiliser phosphorus: implications for their efficient use in agriculture. Proceeding no. 589 of the international fertiliser society. The International Fertiliser Society, York, YO32 5YS, UK

Kokkora MI (2008) Biowaste and vegetable waste compost application to agriculture. PhD Thesis, Cranfield University, UK

Kokkora MI, Hann MJ, Godwin RJ (2008a) The use of vegetable- and bio- waste composts: effects on forage maize production and soil properties. Proceedings of ASABE (American Society of Agricultural and Biological Engineers) Annual International Meeting, 29 June–2 July 2008, Providence, US, ASABE (Paper No.: 083514)

Kokkora MI, Hann MJ, Tyrrel SF (2008b) Organic waste compost parameters in relation to soil properties. Proceedings of EurAgEng (European Society of Agricultural Engineers), International Conference on Agricultural Engineering 'Agricultural and Biosystems Engineering for a Sustainable World', 23 June–25 June 2008, Hersonissos, Crete, Greece (AgEng Paper No.: 2135)

Le MS (2005) Promoting agricultural recycling: factors affecting the assimilation of nutrients from sludge products by crops. A farmer's guide for optimum use of different sludge products (Commercial in confidence). United Utilities North West, Warrington, UK

Le MS, Gedara S (2007) Organo-mineral fertilisers: challenges and opportunities, a state of art review. Draft for discussion (Commercial in confidence). United Utilities North West, Warrington, UK

Lu P, O'Connor GA (2001) Biosolids effects on phosphorous retention and release in some sandy Florida soils. J Environ Qual 30:1059–1063

MAFF (Ministry of Agriculture Fisheries and Food) (1998) Code of good agricultural practice for the protection of water. Booklet PB0587, HMSO, London, UK

MAFF(Ministry of Agriculture Fisheries and Food) (2000) Fertiliser recommendations for agricultural and horticultural crops (RB209), 7th edn. The Stationery Office, London, UK
Maguire RO, Sims JT, Dentel FK, Coale FJ, Mah JT (2001) Relationships between biosolids treatment process and soil phosphorus availability. J Environ Qual 30:1023–1033
Mamo M, Rosen CJ, Halbach TR (1999) Nitrogen availability and leaching from soil amended with municipal solid waste compost. J Environ Qual 28:1074–1082
Marmo L (2008) EU strategies and policies on soil and waste management to offset greenhouse gas emissions. Waste Manage 28:685–689
Martinez J, Burton C, Mallard P (2007) Production, transformation, and utilization of organic fertilisers (manures, composts, organic by-products): What is the scope for their future use? In: De Neve S, Salomez J, Van den Bossche A, Haneklaus S, Van Cleemput O, Hofman G, Schnug E (eds) Proceedings of the 16th international symposium of the international scientific centre of fertilizers (CIEC) 'Mineral versus organic fertilization, conflict or synergism?'. Universiteit Gent, Ghent, Belgium, 16–19 September 2007, pp 14
McGechan MB, Lewis DR (2002) Sorption of phosphorus by soil, Part 1: Principles, equations and models. Biosyst Eng 82(1):1–24
Misra RV, Roy RN, Hiraoka H (2003) On-farm composting methods. FAO (Food and Agriculture Organization), Rome, Italy
MUE (Missouri Univeristy Extension) (2005) Interpretation and recommendation guide for compost. http://www.soiltest.psu.missouri.edu/compostguide.htm. Cited 2005
Mugnier E, Aubain P, Gazzo A, Le Moux J, Brunet H, Landrea B (2001) Disposal and recycling routes for sewage sludge. Report on behalf of DG Environment. Part 3 – Scientific and technical report, Chap 4, pp 39–46 http://europa.eu.int/comm./environment/waste/sludge/pdf/sludge_disposal.htm. Cited 14 August 2007
Nevens F, Reheul D (2003) The application of vegetable, fruit and garden waste (VFG) compost in addition to cattle slurry in a silage maize monoculture: nitrogen availability and use. Eur J Agron 19:189–203
Parkinson RJ, Fuller MP, Groenhof AC (1999) An evaluation of greenwaste compost for the production of forage maize (*Zea mays L.*). Compost Sci Util 7:72–80
Petersen J (2003) Nitrogen fertilizer replacement value of sewage sludge, composted household waste and farmyard manure. J Agric Sci 140:169–182
Rogers BF, Krogmann U, Boyles LS (2001) Nitrogen mineralization rates of soils amended with nontraditional organic wastes. Soil Sci 166:353–363
Skinner RJ, Church BM, Kershaw CD (1992) Recent trends in soil pH and nutrient status in England and Wales. Soil Use Manage 8(1):16–20
Sommers LE, Nelson DW, Yost KJ (1976) Variable nature of chemical composition of sewage sludge. J Environ Qual 5:303–306
Stamatiadis S, Werner M, Buchanan M (1999) Field assessment of soil quality as affected by compost and fertilizer application in a broccoli field (San Benito County, California). Appl Soil Ecol 12:217–225
Sullivan DM, Bary AI, Thomas DR, Fransen SC, Cogger CG (2002) Food waste compost effects on fertilizer nitrogen efficiency, available nitrogen, and tall fescue yield. Soil Sci Soc Am J 66:154–161
Sullivan DM, Cogger CG, Bary AI (2007) Fertilizing with biosolids. Pacific North West Extension Publication 508-E. Oregon State University http://extension.oregonstate.edu/catalog/pdf/pnw/pnw508-e.pdf. Cited 16 September 2008
Svensson K, Odlare M, Pell M (2004) The fertilizing effect of compost and biogas residues from source separated household waste. J Agric Sci 142:461–467
Tang J-C, Inoue Y, Yasuta T, Yoshida S, Katayama A (2003) Chemical and microbial properties of various compost products. Soil Sci Plant Nutr 49(2):273–280
Tisdale SL, Halvin JL, Beaton JD, Nelson WL (1999) Soil fertility and fertilizers: an introduction to nutrient management, 6th edn. Prentice-Hall, New Jersey, US

USEPA (United States Environmental Protection Agency) (1983) Process design manual: land application of municipal sludge. Publication No. EPA-625/1-83-016. USEPA Office of Research and Development, Washington, US

Van-Camp L, Bujarrabal B, Gentile A-R, Jones RJA, Montanarella L, Olazabal C, Selvaradjou S-K (2004) Reports of the technical working groups established under the Thematic Strategy for Soil Protection. EUR 21319 EN/3. Office for Official Publications of the European Communities, Luxembourg

Willett IR, Jakobsen P, Malafant KWJ (1986) Fertiliser and liming value of lime-treated sewage sludge. Fertil Res 8:313–328

Wise SW (1999) A systems approach to the recycling of wastewater sludge phosphorus in agriculture. Thesis Engineering Doctorate, Institute of Agri-Technology, Cranfield University at Silsoe, UK

Withers PJA, Flynn NJ (2006) Application of phosphorus in industrial biosolids to agricultural soils. UKWIR Project SL02B. Interim report prepared for UKWIR (UK Water Industry Research), DEFRA (Department for Environment, Food, and Rural Affairs), and EA (Environment Agency) (In confidence). The United Kingdom Water Industry Research, Queen Anne's Gate, London, UK

Wolkowski RP (2003) Nitrogen management considerations for landspreading municipal solid waste compost. J Environ Qual 32:1844–1850

Zmora-Nahum S, Hadar Y, Chen Y (2007) Physico-chemical properties of commercial composts varying in their source materials and country of origin. Soil Biol Biochem 39:1263–1276

Zwart K (2003) Fate of C and N pools – experience from short and long term compost experiments. In: Applying composts benefits and needs. Federal Ministry of Agriculture, Forestry, Environment and Water Management, Austria, and European Communities, Brussels, pp 77–86

Index

D

M

N

Q

R

Printed by [illegible]
in Hamburg, Germany